T0213531

Analytical Thermodynamics

Dongqing Li

Analytical Thermodynamics

 Springer

Dongqing Li
Mechanical and Mechatronics Engineering
University of Waterloo
Waterloo, ON, Canada

ISBN 978-3-030-90519-4   ISBN 978-3-030-90517-0   (eBook)
https://doi.org/10.1007/978-3-030-90517-0

This Springer imprint is published by the registered company Springer Nature Switzerland AG
The registered company address is: Gewerbestrasse 11, 6330 Cham, Switzerland

Preface

Thermodynamics has a long history. The most well-known founding thermodynamists include: Sadi Carnot, William Thomson, Rudolf Clausius, James Maxwell, Ludwig Boltzmann, and Willard Gibbs, just to name a few. The word "thermodynamics" comes from two Greek words: thermé (means heat) and dynamikos (means force or power). One of the early definitions of thermodynamics is that thermodynamics is the science that deals with relations between heat and work. This definition perhaps reflects the origin of thermodynamics. In the early time, thermodynamics was developed to understand heat engines that absorb heat to produce work. It was in 1850 that Rudolf Julius Clausius (1822~1888) formulated the First Law and the Second Law of Thermodynamics and established the foundation of classical thermodynamics.

Nowadays, thermodynamics is considered as the science that deals with energy, matter, and the laws governing their interactions. Thermodynamics studies the relationships among the properties of macroscopic systems and the restrictions on permissible physical processes. Over the years, thermodynamics has developed into a very important branch of modern science with wide applications not only in engineering but also in physics, in chemistry as well as in life science and in social science.

The theoretical foundation of modern analytical thermodynamics is developed by Josiah Willard Gibbs (1839–1903). Gibbs was a theoretical physicist, chemist, and mathematician. He was awarded the first American Ph.D. in engineering in 1863 at Yale University. In 1901, Gibbs was granted the highest honor of his time, the Copley Medal of the Royal Society of London, for being "the first to apply the second law of thermodynamics to the exhaustive discussion of the relation between chemical, electrical, and thermal energy and capacity for external work." I still remember the excitement I had many years ago when I first time read the book "*The Scientific Papers of J. Willard Gibbs, Volume One: Thermodynamics.* Woodbridge, Conn.: Ox Bow Pr., 1970". At that time, I was a Ph.D. student at the University of Toronto. I found this book at the basement of one of the university's libraries. Immediately I was amazed by the elegance and power of the theoretical structure of the analytical thermodynamics presented in that book.

Why is the analytical thermodynamics important? As we know, the foundation of thermodynamics is several laws. Although thermodynamics' laws cannot be proven, they are correct, because they result from our observation. These laws are facts, the summarized facts and the correctly generalized facts. All thermodynamic theories are the logical development of these observation and generalization. That is, all thermodynamics theories are the logical development of the thermodynamic laws. The importance of analytical thermodynamics is the fact that the logical structure of analytical thermodynamics has made it possible for us to predict the behavior of the real world.

I have taught advanced thermodynamics courses to engineering graduate students in the University of Alberta (Canada), the University of Toronto (Canada), Vanderbilt University (USA), and University of Waterloo (Canada) for 30 years in my academic career. Unfortunately, there are few textbooks of thermodynamics at graduate level. This book represents a major part of the lecture notes that I have developed over years. My philosophy is to make thermodynamics as easy to understand as possible. Before taking my course, many students did not know that thermodynamics has such an interesting side and is easy to learn (without using property tables and charts, just modeling and analyzing). I sincerely hope that this book will be useful to professors teaching advanced thermodynamics and to graduate students learning thermodynamics.

This book is designed for an advanced thermodynamics course. Chapter 1 will introduce the theory of analytical thermodynamics by using the postulation approach. The content of Chaps. 2–4 are novel and unique. Chapters 2 and 3 serve two purposes. One is to expand the horizon from what has been presented in Chap. 1. The other purpose is to demonstrate how to apply the general principles of analytical thermodynamics to model and to solve problems by studying special systems under external fields and systems of surfaces or interfaces. Chapter 4 provides an analytical understanding of the second law of engineering thermodynamics.

It is very fortunate that I have the opportunity to learn and teach thermodynamics in my life. Writing this book also brings me a lot of enjoyment. I wish everyone who reads this book would appreciate it too.

Waterloo, Canada   Dongqing Li

References

H. B. Callen, (1985). *Thermodynamics and an introduction to thermostatistics* (2nd ed.). Hoboken NY: John Wiley & Sons. ISBN 9780471862567.

J. Willard Gibbs (1970). *The scientific papers of J. Willard Gibbs. 1: Thermodynamics*. Publisher: Woodbridge, Conn.: Ox Bow Pr.

W. Gibbs (1993). *The scientific papers of J. Willard Gibbs, volume one: Thermodynamics*. Ox Bow Press. ISBN 978-0-918024-77-0. OCLC 27974820.

Contents

1 Basics of Analytical Thermodynamics 1
   1.1   Definitions 1
   1.2   Postulates 4
   1.3   Fundamental Equations 6
   1.4   Euler Equation and Gibbs–Duhem Equation 9
   1.5   Simple Equilibrium 14
   1.6   Extreme Principles of Equilibrium States 21
   1.7   Legendre Transformations and Thermodynamic Potentials 23
   1.8   Minimum Principles of Thermodynamic Potentials/Free
        Energies 30
   1.9   Applications of Minimum Principle of Thermodynamic
        Potentials 35
   1.10 Maxwell Relations 39
   1.11 Thermodynamic Characteristics of Dielectric Media 46
   1.12 Introduction to Thermodynamic Stability 49
   1.13 Phase Change and Clapeyron Equation 53
   1.14 Chemical Potentials 57
   1.15 Boiling Temperature and Freezing Temperature of Dilute
        Solutions 65
   1.16 Gibbs Phase Rule 69
   1.17 Introduction to High-Order Phase Change 72

2 Modelling Homogeneous and Heterogeneous Systems 75
   2.1   Simple Elastic Solid 76
   2.2   Simple Electrolyte Solution Systems 80
   2.3   Systems in Gravitational Field and in Centrifugal Field 88

3 Thermodynamics of Interfaces and Three-Phase Contact Lines 101
   3.1   Introduction to Interfaces and Three-Phase Contact Lines 101
   3.2   Thermodynamics of Surfaces 104
   3.3   Thermodynamics of Three-Phase Contact Lines 109
   3.4   Equilibrium Conditions of Droplets and Bubbles 111

3.5   Equilibrium Conditions of Sessile Drops   116
3.6   From Laplace Equation to Capillary Rise and Meniscus
        Shape   128
3.7   Curvature Effects on Equilibrium Pressure and Temperature   137
3.8   Solute Effect on Equilibrium Radius of Droplets   142
3.9   Heterogeneous Nucleation   145
3.10   Equilibrium Condition of a Bubble in a Uniform Electric
        Field   152
3.11   Effects of Applied Electrical Field on Contact Angles   156
3.12   Effects of Electric Double Layer on Contact Angle   160
3.13   Modelling Surface Processes by Using Surface Free Energy   164

4 **Second Law of Engineering Thermodynamics**   177
4.1   Irreversible Processes   178
4.2   Limitation of First Law of Thermodynamics   180
4.3   Second Law and Equations   182
4.4   The 2nd Law Requirement on Thermal Cycles   186
4.5   Applying Second Law Equation to a Human Body   191

About the Author

Dr. Dongqing Li was a professor in the Department of Mechanical Engineering, University of Alberta from 1993 to 2000. He later joined the Department of Mechanical and Industrial Engineering, University of Toronto, in 2000 as a tenured full professor. From October 2005 to August 2008, he was the H. Fort Flowers Chair Professor of Mechanical Engineering at Vanderbilt University, USA. Since September 2008, he has joined the University of Waterloo as a tenured full professor, and he was the Canada Research Chair in Microfluidics and Nanofluidics from 2008 to 2013.

He is an internationally leading expert in the area of microfluidics and nanofluidics. He was the Editor-in-Chief of an international journal—*Microfluidics and Nanofluidics* (Springer) from 2004 to 2012. He was also the Editor-in-Chief of the *Encyclopedia of Microfluidics and Nanofluidics* (1st edition and 2nd edition) (Springer). He has published over 330 papers in top international journals, many book chapters, and three books. The Google Scholar citation number of his publication is over 26700. The H-index of him is 78.
(https://scholar.google.ca/citations?user=F1dBly4AAAAJ&hl=en).

Examples of books published by Dr. Dongqing Li:
Dongqing Li, "Electrokinetics in Microfluidics", Academic Press, London, 2004.
Dongqing Li, Editor-in-Chief, "Encyclopedia of Microfluidics and Nanofluidics", Springer, 1st edition, 2008. 2nd edition, 2015.

Chapter 1
Basics of Analytical Thermodynamics

Abstract This chapter will introduce the theory and analytical methods of thermodynamics. The key are the four postulates which are the foundation of the analytical thermodynamics. Then the fundamental equations, Euler equation and Gibbs-Duhem equation are introduced. A fundamental equation contains all thermodynamic information of a given system. Several thermodynamic potential functions are introduced by using Legendre transformation. Knowing the fundamental equation, thermodynamic equilibrium conditions can be readily determined by applying the entropy maximum principle or the energy minimum principle. Several explicit forms of the chemical potentials for simple and idealized systems are introduced. They are very useful tools in determining the equilibrium conditions of coexisting phases. As examples, the boiling point and the freezing point of ideal (dilute) solutions are analyzed.

1.1 Definitions

System and surrounding: A thermodynamic system is defined as a quantity of matter or a region in space that is chosen for study. In other words, **the system is what we want to study**. The region outside the system is called the **surroundings** (i.e., Surroundings are physical space outside the system boundary). The real or imaginary surface that separates the system from its surroundings is called the **boundary**. The boundary of a system may be fixed or movable or deformable.

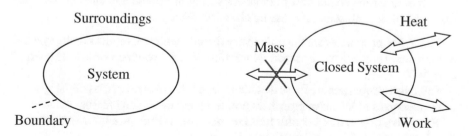

Systems can be further classified into several groups:

© The Author(s), under exclusive license to Springer Nature Switzerland AG 2022
D. Li, *Analytical Thermodynamics*,
https://doi.org/10.1007/978-3-030-90517-0_1

A **closed system** consists of a **fixed** amount of **mass** and no mass may transfer across the system boundary. A closed system may exchange energy such as heat and work with surroundings.

An **open system** is a system that has **mass** as well as energy **transfer across the system's boundary**.

An **isolated system** is a system without any interactions with the surroundings (i.e., no mass, heat or work may cross the boundaries).

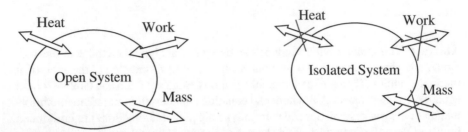

Simple system: A simple system is a system that is homogeneous, isotropic, uncharged, not subject to field interactions (e.g. electric, or gravitational fields), and has no surface or boundary effects. In this book, all systems are considered as simple systems unless specified otherwise.

Internal energy: Internal energy, U, of a thermodynamic system includes the intrinsic energy of individual molecules, the energy of the random motion of molecules, and the energy of molecular interactions. The internal energy U does not include any kinetic energy and potential energy of the system.

Thermodynamic equilibrium state: A thermodynamic equilibrium state is a state in which: (1) all properties are independent of time and are uniform everywhere in the system; (2) the system maintains thermal, mechanical, phase and chemical equilibrium.

More specifically, the system is in

(a)   thermal equilibrium if it has constant and uniform temperature T.
(b)   mechanical equilibrium if it has constant and uniform pressure P.
(c)   phase equilibrium if it has no net phase change. For example, for a liquid-vapor two phase system, the amounts of liquid water and water vapor remain the same; and no net flow of molecules from one phase into another.
(d)   chemical equilibrium if it has no chemical reactions.

Extensive properties are those that vary directly with size or mass of the system. Some examples of extensive properties are: mass m, total volume V, and total internal energy U.

Intensive properties are those that are independent of size or mass of the system. Some examples of intensive properties are: temperature T, and pressure P.

Extensive properties per unit mass or per unit volume are intensive properties. For example, the specific volume v is defined as

$$v = \frac{volume}{mass} = \frac{V}{m} \quad (\text{m}^3/\text{kg})$$

and density ρ is defined as

$$\rho = \frac{mass}{volume} = \frac{m}{V} \quad (\text{kg/m}^3)$$

They are intensive properties.

Note $v = 1/\rho$.

What is a mole? Mole is a unit for expressing a quantity of a substance. A mole of a substance is the amount that contains 6.022×10^{23} particles (e.g., atoms, molecules, ions) of that substance.

Mole number is the number of moles, and is defined as:

$$N = \frac{mass}{molar\ mass} = \frac{m}{M}$$

where M is called the molar mass or molecular weight.

For example, $M_{H_2} = 2.0 \frac{\text{kg}}{\text{kmol}}$, $M_{air} = 28.97 \frac{\text{kg}}{\text{kmol}}$.

That is, 1 kmol (1 kmol = 1000 mol) of air has a mass of 28.97 kg.

Note: $m = N M$.

Saturation state means the state when phase change starts. For example, a saturated liquid is a liquid that is about to vaporize. A saturated vapor is a vapor that is about to condense into liquid.

At a given pressure, the temperature at which a pure substance starts changing its phase is called the **saturation temperature**. For example, at 1 atm, water's saturation temperature is 100 °C.

Process: Any change from one state to another is called a process.

Thermal Cycle: One or a series of connected processes with identical end states is called a cycle. The Figure below is an example of a cycle composed of three processes. The properties of the system at the end of the cycle are the same as at its initial state.

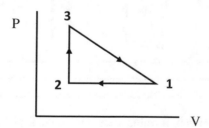

Heat (Thermal) Reservoir: A heat reservoir is a system that has infinite amount of thermal energy (heat) and a constant temperature. A finite amount of heat transfer into or from the heat reservoir will not change its temperature.

A higher temperature heat reservoir from which heat is transferred is often called **a heat source**. A lower temperature heat reservoir to which heat is transferred is called **a heat sink**.

Heat Engine: A heat engine is a thermal cycle system. During the cycle, it receives heat from a heat source, converts part of the heat into work and rejects the rest of the heat to a heat sink.

Heat Pump (or refrigerator)

A heat pump or a refrigerator is a thermal cycle system that removes heat from a low temperature body and delivers heat to a high temperature body. To do so, the heat pump or refrigerator requires external work input.

1.2 Postulates

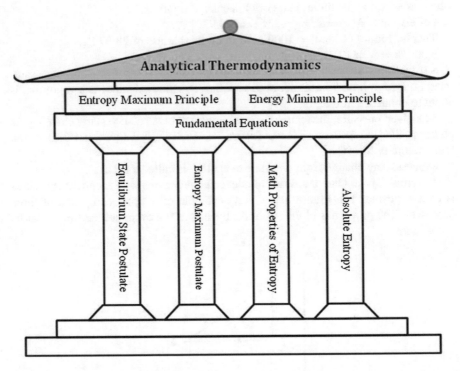

There are four postulates forming four fundamental pillars in the structure of analytical thermodynamics theory. The entire analytical thermodynamics theory and applications are the logic derivation of these four postulates.

I.   **Equilibrium State Postulate**

The equilibrium states of a simple system with r independent chemical species are completely characterized by $(r + 2)$ **extensive variables**: internal energy U, volume V, and mole numbers, $N_1, N_2, \ldots N_r$, of the independent chemical species. In other words, the equilibrium states of a simple system with r independent chemical species are completely characterized by $(r + 2)$ **extensive variables,** $(U, V, N_1, \ldots N_r)$.

II. Entropy Maximum Postulate

For a simple thermodynamic system, there exists a function of extensive variables, called entropy S,

$$S = S(U, V, N_1, N_2 \ldots N_r).$$

If the system is an isolated system, entropy S assumes its maximum value at an equilibrium state. That is,

$$S = S_{\max} \quad \text{at equilibrium.}$$

Mathematically, $S = S_{\max}$ at equilibrium requires

$$\begin{cases} \frac{\partial S}{\partial X_i} = 0 & (X_i = U, V, N_1, \ldots, N_r) \\ \delta S < 0, \quad \frac{\partial^2 S}{\partial X_i \partial X_j} < 0 \end{cases}$$

In the above, the symbol δ means small variation from the maximum point.

III. Mathematical Properties of Entropy:

(1) The total entropy of a composite system is the sum of the entropies of all the sub-systems.

$$S = \sum_i S_i$$

(2) Entropy function is continuous, differentiable and is a monotonically increasing function of the energy. Mathematically these require satisfying the following conditions:

$$\left(\frac{\partial^n S}{\partial X_i^n} \right)_{X_j} \quad \text{exists for } n = 1, 2, \ldots$$

$$\left(\frac{\partial S}{\partial U} \right)_{V, N_1, \ldots N_r} > 0$$

(3)   For a simple system, entropy S is a first-order homogeneous function of the $(r + 2)$ extensive variables. In mathematical language, this is expressed as:

$$S(\lambda U, \lambda V, \lambda N_1, \ldots \lambda N_r) = \lambda S(U, V, N_1, \ldots N_r)$$

where λ is an arbitrary constant. The above first order of homogeneous function implies that if the size of the system increases by λ times, the total entropy of the system will also increase by λ times.

(4)   The above mentioned mathematics properties ensure the existence of the following functions.

$$S = S(U, V, N_1, \ldots N_r) \quad \text{and} \quad U = U(S, V, N_1, \ldots N_r).$$

These two equations are the **thermodynamic fundamental equations of a simple system in entropy form and in energy form, respectively.**

IV.   **Absolute Entropy**

$$S = 0 \quad \text{at} \quad T = 0 \text{ K}.$$

This postulate provides a base to calculate the absolute entropy at a given state.

1.3 Fundamental Equations

Thermodynamic fundamental equations contain all the thermodynamic information of a system. Knowing the fundamental equation of a system, the equilibrium conditions of this system can be derived analytically, as will be shown in later sections. In the form of internal energy, the fundamental equation is given by:

$$U = U(S, V, N_1, \ldots N_r)$$

If we use X_i ($i = 1, 2, \ldots, n = r + 2$) to represent the extensive variables, S, V, and N_i, the above fundamental equation can be written as:

$$U = U(X_1, X_2, \ldots X_n) = U(\{X_i\})$$

The differential form of the fundamental equation is

$$dU = \sum Y_i dX_i$$

where

$$Y_i = \left(\frac{\partial U}{\partial X_i}\right)$$

is **the intensive variable** corresponding to the extensive variable X_i. More specifically, consider a 3-D bulk phase as an example,

$$U = U(S, V, N_1, \ldots N_r)$$

$$dU = \left(\frac{\partial U}{\partial S}\right)_{V,N_i} dS + \left(\frac{\partial U}{\partial V}\right)_{S,N_i} dV + \sum_i \left(\frac{\partial U}{\partial N_i}\right)_{S,V,N_{j\neq i}} dN_i$$

$$= TdS - PdV + \sum_i \mu_i\, dN_i$$

where the intensive variables are defined as

$$T = \left(\frac{\partial U}{\partial S}\right)_{V,N_i}, \qquad -P = \left(\frac{\partial U}{\partial V}\right)_{S,N_i}, \qquad \mu_i = \left(\frac{\partial U}{\partial N_i}\right)_{S,V,N_{j\neq i}}$$

In the above, T is the **temperature**, P is the **pressure**, and μ_i is the **chemical potential**. The three terms in the differential form of the fundamental equation represent thermal energy (TdS), mechanical work (PdV) and chemical work ($\mu_i\, dN_i$), respectively.

Similarly, for the entropy form of the fundamental equation, we have

$$S = S(U, V, N_1, \ldots N_r)$$

$$dS = \left(\frac{\partial S}{\partial U}\right)_{V,N_i} dU + \left(\frac{\partial S}{\partial V}\right)_{U,N_i} dV + \sum_i \left(\frac{\partial S}{\partial N_i}\right)_{U,V,N_{j\neq i}} dN_i$$

$$= \frac{1}{T} dU + \frac{P}{T} dV - \sum_i \frac{\mu_i}{T} dN_i$$

where

$$\frac{1}{T} = \left(\frac{\partial S}{\partial U}\right)_{V,N_i}, \qquad \frac{P}{T} = \left(\frac{\partial S}{\partial V}\right)_{U,N_i}, \qquad -\frac{\mu_i}{T} = \left(\frac{\partial S}{\partial N_i}\right)_{U,V,N_{j\neq i}}$$

Example 1: Is the following function

$$S = A(NVU)^{1/3}$$

a fundamental equation? In this equation A is a constant.

Solution

If this function is a fundamental equation, it must satisfy Postulate III and Postulate IV. Let examine the given function against the conditions specified in Postulate III and Postulate IV.

(1)   Yes, this function is a continuous, differentiable function of N, V, and U. All orders of partial derivatives exist.

(2) $\left(\frac{\partial S}{\partial U}\right)_{V,N} = \frac{A}{3} \frac{[NV]^{1/3}}{U^{2/3}} > 0$

That is, the is function S is a monotonically increasing function of energy U.

(3) $\lambda S = A[(\lambda N)(\lambda V)(\lambda U)]^{\frac{1}{3}} = \lambda A(NVU)^{1/3}$

It is a 1st-order homogeneous function.

(4)   From the given function, we have

$$U = \frac{S^3}{A^3 NV} \quad T = \left(\frac{\partial U}{\partial S}\right)_{V,N} = \frac{3S^2}{A^3 NV}$$

Clearly, when T → 0, S → 0.

Conclusion: This function satisfies all conditions in Postulate III and Postulate IV. It is a fundamental equation.

Example 2: For a given function

$$S = A(NU/V)^{\frac{2}{3}}$$

where A is a constant, is it a fundamental equation?

Solution

Because

$$\lambda S = A[(\lambda N)(\lambda U)/(\lambda V)]^{\frac{2}{3}}$$

$$A[(\lambda N)(\lambda U)/(\lambda V)]^{\frac{2}{3}} = A(\lambda)^{\frac{2}{3}}(NU/V)^{\frac{2}{3}} \neq \lambda A(NU/V)^{\frac{2}{3}}$$

it is not a first-order homogeneous function. Therefore, it is not a fundamental equation.

Example 3: For a given function

$$S = AV^3/(NU)$$

where A is a constant, is it a fundamental equation?

Because entropy is a monotonically increasing function of energy, that is,

$$\left(\frac{\partial S}{\partial U}\right)_{V,N} > 0$$

However, for this given function,

$$\left(\frac{\partial S}{\partial U}\right)_{V,N} = -\frac{AV^3}{NU^2} < 0$$

Therefore, it is not a fundamental equation.

Home Works

(1) Determine if the following equation is acceptable as a fundamental equation.

$$U = A N V[1 + S/(NR)]\exp(-S/NR)$$

where A and R are constant.

(2) For a system with r independent chemical components, show $du = Tds - Pdv + \sum_{j=1}^{r-1} (\mu_j - \mu_r)dx_j$ where $x_j = \frac{N_j}{N}$

1.4 Euler Equation and Gibbs–Duhem Equation

As introduced in the last section, a fundamental equation is a first-order homogeneous function. By definition,

$$U(\lambda X_1 \lambda X_2 \ldots \lambda X_n) = \lambda U(X_1, X_2 \ldots X_n)$$

where λ is an arbitrary real number and can be any value, and X_i $(i = 1, 2, \ldots n)$ are the **extensive variables** (i.e., S, V, $N_1 \ldots N_r$). For convenience, the above equation is re-written as:

$$U(\{\lambda X_i\}) = \lambda U(\{X_i\})$$

Differentiating the above equation with respect to λ gives:

$$\sum_i \frac{\partial U(\{\lambda X_i\})}{\partial(\lambda X_i)} \frac{\partial(\lambda X_i)}{\partial \lambda} = U(\{X_i\})$$

$$\sum \frac{\partial U(\{\lambda X_i\})}{\partial(\lambda X_i)} X_i = U(\{X_i\})$$

Since λ is an arbitrary real number, if we choose $\lambda = 1$, the above equation becomes

$$\sum \frac{\partial U(\{X_i\})}{\partial (X_i)} X_i = U(\{X_i\})$$

Recall the **intensive parameter** is defined as

$$Y_i = \left(\frac{\partial U}{\partial X_i} \right),$$

it follows that

$$\sum_i Y_i X_i = U(\{X_i\}).$$

The above equation can be rearranged as:

$$U(\{X_i\}) = \sum_i Y_i X_i$$

This is the so-called **Euler equation**.

Let us see what the Euler equation means by the following example. Let us consider a simple 3-D bulk phase system. The energy form of the fundamental equation is given by

$$U = U(S, V, N_1, \ldots N_r)$$

The extensive variables in the energy form of the fundamental equation are:

$$S, V, N_1, \ldots N_r,$$

As defined in the last section, the corresponding intensive parameters are:

$$T = \left(\frac{\partial U}{\partial S} \right)_{V,N_i}, \qquad -P = \left(\frac{\partial U}{\partial V} \right)_{S,N_i}, \qquad \mu_i = \left(\frac{\partial U}{\partial N_i} \right)_{S,V,N_j \neq i}$$

According to the Euler equation, we will have

$$U = \sum_i Y_i X_i$$

where X_i is an extensive variable in the energy function U (i.e., S, V, N_i), and Y_i is the intensive parameter corresponding to X_i (i.e., T, P, μ_i). Therefore, the specific form of the Euler equation for this case is:

$$U = TS - PV + \sum_i \mu_i N_i$$

Clearly, Euler equation provides us with an explicit form of the internal energy function, relating U with S, V, N_i, T, P and μ_i.

Similarly, we can prove that the Euler equation for the entropy form of the fundamental equation is given by:

$$S(\{X_i\}) = \sum_i Y_i X_i$$

where X_i is an extensive variable in the entropy function (i.e., U, V, N_i), and Y_i is the intensive parameter corresponding to X_i.

More specifically, for a simple 3-D bulk phase system,

$$S = S(U, V, N_1, \ldots N_r),$$

The extensive variables in the entropy form of the fundamental equation are:

$$U, V, N_1, \ldots N_r,$$

As defined in the last section, the corresponding intensive parameters are:

$$\frac{1}{T} = \left(\frac{\partial S}{\partial U}\right)_{V,N_i}, \quad \frac{P}{T} = \left(\frac{\partial S}{\partial V}\right)_{U,N_i}, \quad -\frac{\mu_i}{T} = \left(\frac{\partial S}{\partial N_i}\right)_{U,V,N_j \neq i}$$

According to the Euler equation for the entropy form of the fundamental equation:

$$S(\{X_i\}) = \sum_i Y_i X_i$$

It follows that:

$$S = \frac{U}{T} + \frac{P}{T}V - \sum_i \frac{\mu_i}{T} N_i$$

Clearly, Euler equation provides us with an explicit form of the entropy function (not only S is a function of U, V and mole numbers N_i).

Gibbs–Duhem Relation

By differentiating the Euler equation, $U = \sum Y_i X_i$, we can have

$$dU = \sum_i Y_i \, dX_i + \sum_i X_i \, dY_i$$

Also, the differential form of the fundamental equation gives us:

$$dU = \sum_i Y_i\, dX_i$$

Combining the above two equations yields

$$\sum_i X_i\, dY_i = 0$$

where X_i is an extensive variable in the energy function U (i.e., S, V, N_i), and Y_i is the intensive parameter corresponding to X_i (i.e., T, P, μ_i). This equation is called the **Gibbs–Duhem equation**. As seen from the above equation, Gibbs–Duhem equation provides a correlation among intensive parameters, Y_i.

For a simple, bulk phase system,

$$U = U(\{X_i\}) = U(S, V, N_1, \ldots N_r)$$

and

$$dU = \sum Y_i dX_i = TdS - PdV + \sum \mu_i dN_i$$

The Gibbs–Duhem equation for such a system is given by:

$$\sum_i X_i\, dY_i = 0$$

That is,

$$SdT - VdP + \Sigma\, N_i\, d\,\mu_i = 0.$$

This is an important equation. For a single component system, $r = 1$, the Gibbs–Duhem equation becomes:

$$SdT - V\,dP + Nd\mu = 0$$

Dividing both sides of this equation by the mole number leads to:

$$d\mu = \tfrac{1}{N}(-SdT + V\,dP)$$

Clearly, it indicates that $\mu = \mu\,(T, P)$.
If T = constant and P = constant, from

$$SdT - V\,dP + \sum N_i d\mu_i = 0$$

We have

$$\sum N_i d\mu_i = 0$$

For a binary system, e.g., a solution consisting of component 1 and component 2, the above equation gives:

$$N_1 d\mu_1 + N_2 d\mu_2 = 0$$

or

$$d\mu_1 = -\frac{N_2}{N_1} d\mu_2$$

This equation shows how chemical potentials of different components are related to each other.

Example of simple systems

For an ideal gas of a single component ($r = 1$), we know

$$\left. \begin{array}{l} PV = NRT \\ U = CNRT \end{array} \right\} \quad or \quad \left\{ \begin{array}{l} Pv = RT \\ u = CRT \end{array} \right.$$

We would like to find the expressions for the specific entropy s (i.e., on per unit mole basis) and the chemical potential μ.

According to the entropy form of the fundamental equation, we have

$$ds = \frac{1}{T} du + \frac{P}{T} dv$$

In the above, we have considered that the mass of the ideal gas system is one kilo-mole and fixed. Therefore, the fundamental equation does not have the ($\mu \, dN$) term, and entropy, internal energy and volume are on per unit mole basis (called specific properties, and in small letters). Using the given relations yields:

$$ds = \frac{CR}{u} du + \frac{R}{v} dv$$

Integration yields

$$s = s_o + CR \ln\left(\frac{u}{u_o}\right) + R \ln\left(\frac{v}{v_o}\right) = s_o + CR \ln\left(\frac{T}{T_o}\right) + R \ln\left(\frac{TP_o}{T_o P}\right)$$

Recall that a fundamental equation is a function of extensive variables. The above equation is not a fundamental equation because it is expressed in terms of intensive variables. Using the given conditions,

$$PV = NRT \quad \text{and} \quad U = CNRT$$

We can find correlations between T and U, P and UV. If we replace T/T_0 by U/U_0, and P_0/P by $(U_0V)/(UV_0)$ in the above equation, and multiply both sides of the equation by mole number N, then we can have an entropy form of fundamental equation.

The expression of the chemical potential μ can be derived by using the Euler equation and the derived expression for s, i.e.,

$$\mu = u + Pv - Ts$$

$$= CRT + RT - Ts_o - TCR \ln\left(\frac{T}{T_o}\right) - TR \ln\left(\frac{TP_o}{T_0 P}\right)$$

$$= \phi(T) + RT \ln P$$

Home work

A particular system has the following two equations of state:

　　$T = 3A\, s^2/v$, and $P = As^3/v^2$ where A is a constant.

Find (a) μ as a function of s and v (hint: Gibbs–Duhem equation), (b) the fundamental equation of the system (hint: Euler equation).

1.5　Simple Equilibrium

As we have introduced in the first section of this chapter, a thermodynamic equilibrium state is a state in which: (1) All properties are independent of time and are uniform everywhere in the system; (2) The system maintains thermal, mechanical, phase and chemical equilibrium. In this book, we will not discuss equilibrium states of chemical reactions. The focus will be thermal equilibrium, mechanical equilibrium and the equilibrium of phase change and mass exchange.

Thermal Equilibrium

Consider an isolated system consisting of subsystems 1 and 2, as shown in the figure below.

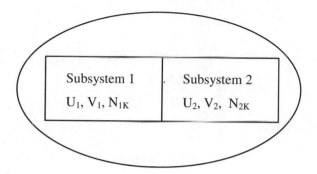

The system is subject to the following constraints.

External constraints

$$U_1 + U_2 = \text{constant}$$
$$N_{1k} + N_{2k} = \text{constant}, \, k = 1, 2, \ldots r$$
$$V_1 + V_2 = \text{constant}$$

Internal constraints

$$N_{1k} = \text{constant} \quad N_{2k} = \text{constant} \quad k = 1, 2, \ldots r$$
$$V_1 = \text{constant} \quad V_2 = \text{constant}$$

Comparing the above two sets of constraints, we see that all parameters are fixed, except U_1 and U_2. This means that energy exchange between the two subsystems is allowed.

According to the **Entropy Maximum Postulate**, the total entropy of the system is maximum at equilibrium, i.e.,

$$S = S_1 + S_2 = \text{maximum at equilibrium}$$

Mathematically, it requires

$$dS = d(S_1 + S_2) = dS_1 + dS_2 = 0.$$

Because

$$dS = \frac{1}{T}dU + \frac{P}{T}dV - \Sigma \frac{\mu_k}{T}dN_k$$

and from the constraints, we have

$$dN_k = 0, dV = 0.$$

Therefore,

$$dS_1 = \frac{dU_1}{T_1}, \quad \text{and} \quad dS_2 = \frac{dU_2}{T_2},$$

$$dS = \frac{dU_1}{T_1} + \frac{dU_2}{t_2}$$

Using the constraint:

$$U_1 + U_2 = \text{constant},$$

we have

$$dU_1 = -dU_2,$$

Therefore,

$$dS = dU_2\left(\frac{1}{T_2} - \frac{1}{T_1}\right) = 0$$

This requires

$$T_1 = T_2 = T$$

This is the thermal equilibrium condition, i.e., all systems in a mutual thermal equilibrium have the same temperature. From the above analysis, we see that **as long as the energy exchange between the subsystems is allowed, i.e., $dU_1 = -dU_2$, the temperatures of the subsystems must be equal at the final equilibrium state**.

The heat flow direction: Because the heat transfer is an irreversible process which generates entropy, $dS > 0$. Therefore, according to the dS equation:

$$dS = dU_2\left(\frac{1}{T_2} - \frac{1}{T_1}\right) > 0$$

If $T_1 > T_2$, it requires $dU_2 > 0$. This means the energy (heat) transfers from subsystem 1 to subsystem 2. If $T_1 < T_2$, it requires $dU_2 < 0$. This means the energy (heat) transfers from subsystem 2 to subsystem 1. In other words, heat always transfers from a higher temperature place to a lower temperature place.

Mechanical Equilibrium

Consider an isolated system comprised of two subsystems 1 and 2, and the partition separating the two subsystems can conduct heat and is movable.

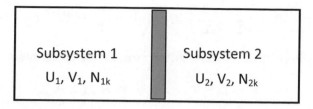

This isolated system is subject to the following constraints.

External constraints

$$U_1 + U_2 = \text{constant}$$
$$N_{1k} + N_{2k} = \text{constant}, k = 1, 2, \ldots r$$
$$V_1 + V_2 = \text{constant}$$

Internal constraints

$$N_{1k} = \text{constant}, \qquad N_{2k} = \text{constant} \qquad k = 1, 2, \ldots r.$$

Comparing the above two sets of constraints, we see that only mass exchange between the two subsystems is not allowed.

The entropy maximum postulate requires the following, at an equilibrium state,

$$dS = d(S_1 + S_2) = dS_1 + dS_2 = 0$$

We know

$$dS = \frac{1}{T}dU + \frac{P}{T}dV - \Sigma\frac{\mu_k}{T}dN_k$$

According to the constraints,

$$dN_{1k} = 0 \text{ and } dN_{2k} = 0,$$

it follows

$$dS = \frac{dU_1}{T_1} + \frac{P_1}{T_1}dV_1 + \frac{dU_2}{T_2} + \frac{P_2}{T_2}dV_2$$

Using the constraints, we have

$$dU_1 = -dU_2, \text{ and } dV_1 = -dV_2.$$

Thus,

$$dS = \left(\frac{1}{T_2} - \frac{1}{T_1}\right) dU_2 + \left(\frac{P_2}{T_2} - \frac{P_1}{T_1}\right) dV_2 = 0.$$

Since dU_2 and dV_2 are arbitrary and independent, we conclude that

$$T_1 = T_2 = T \quad \text{and} \quad P_1 = P_2 = P.$$

Here

$$P_1 = P_2 = P$$

is the **mechanical equilibrium condition**. From the above derivation, we see that, **as long as energy exchange and volume exchange between subsystems are allowed, the temperatures and pressures of the subsystems must be equal at equilibrium.**

Mass Exchange Equilibrium

Consider an isolated system as shown in the figure below. The system is separated by a partition into two chambers (subsystems); each contains a number of different molecules. Assume the partition is a membrane that allows only type 1 molecules to pass through it.

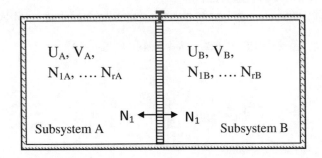

This composite system is subject to the following constraints.

External constraints

$$U_A + U_B = \text{constant}.$$

$$V_A + V_B = \text{constant}$$

$$N_{iA} + N_{iB} = \text{constant}. \quad i = 1, 2, \ldots r$$

Internal constraints

$$V_A = \text{constant}, V_B = \text{constant}$$

$$N_{1A} + N_{1B} = \text{constant}.$$

(i.e., only type 1 molecules are allowed to exchange between the two subsystems)

$$\left.\begin{array}{l} N_{iA} = \text{constant.} \\ N_{iB} = \text{constant} \end{array}\right\} \quad i = 2, 3, \ldots r$$

Because the total entropy of the composite system is maximum at equilibrium, i.e.,

$$dS = dS_A + dS_B = 0,$$

and

$$dS = \frac{1}{T}dU + \frac{P}{T}dV - \Sigma \frac{\mu_k}{T}dN_k$$

and using the constraints of constant volume and constant mass,

$$dV = 0, \quad \text{and} \quad dN_k = 0 \ (k = 2, 3, \ldots r)$$

we have

$$dS = \frac{dU_A}{T_A} - \frac{\mu_{1A}}{T_A}dN_{1A} + \frac{dU_B}{T_B} - \frac{\mu_{1B}}{T_B}dN_{1B} = 0$$

Furthermore, the constraints of

$$U_A + U_B = \text{constant}$$

and

$$N_{1A} + N_{1B} = \text{constant}$$

lead to

$$dU_A = -dU_B \quad \text{and} \quad dN_{1A} = -dN_{1B}$$

Therefore,

$$dS = \left[\frac{1}{T_B} - \frac{1}{T_A}\right]dU_B + \left[\frac{\mu_{1A}}{T_A} - \frac{\mu_{1B}}{T_B}\right]dN_{1B} = 0.$$

It requires:

$$T_B = T_A = T \quad \text{and} \quad \mu_{1A} = \mu_{1B}$$

Here

$$\mu_{1A} = \mu_{1B}$$

is the **equilibrium condition of mass exchange**. From the above derivation, we see that **as long as the mass (molecules) exchange between two systems is allowed, the chemical potential of those molecules in these two systems must be equal at equilibrium.**

Let us consider the **mass transfer direction**. Assume $T_A = T_B = T$. Thus,

$$dS = \frac{\mu_{1A} - \mu_{1B}}{T} dN_{1B}.$$

For an isolated system, because the mass transfer is an irreversible process which generates entropy, i.e., $dS > 0$. Therefore, we see the following possibilities:

If $\mu_{1A} > \mu_{1B}$, $\rightarrow dN_{1B} > 0$, i.e., mass transfers from A to B.
If $\mu_{1A} < \mu_{1B}$, $\rightarrow dN_{1B} < 0$, i.e., mass transfers from B to A.

The above analysis shows that mass transfers from a place with a higher chemical potential to a place with a lower chemical potential. That is, the chemical potential difference is the driving force in mass transfer, just like the temperature difference in energy (heat) transfer.

Home Work

A thin glass tube contains two liquid phases and one vapor phase, as illustrated in the figure below. Liquid 1 is a pure liquid; liquid 2 is a different pure liquid; the vapor phase contains both components (i.e., vapor of liquid 1 and vapor of liquid 2). Consider the glass tube with these three fluid phases as an isolated system. Derive the thermodynamic equilibrium conditions by using the entropy maximum principle.

| Liquid 1 | vapor | Liquid 2 |

1.6 Extreme Principles of Equilibrium States

From the previous examples of finding equilibrium conditions, we see that the maximum entropy postulate leads to a general principle, the **Entropy Maximum Principle**: **For a simple, isolated system, the equilibrium value of any unconstrained internal parameters will maximize the entropy under the given value of the total internal energy**.

The "unconstrained internal parameters" in the above statement are the parameters that are allowed to exchange between subsystems. For example, in the last section, the internal energy U_1 and U_2 are the unconstrained internal parameters in the first case of finding thermal equilibrium. In the left figure below, because the piston is free to move, the volumes V_1 and V_2 are the unconstrained internal parameters. In the right figure below, type-1 molecules are allowed to exchange across the membrane partition. The mole numbers N_{1A} and N_{1B} are the unconstrained internal parameters.

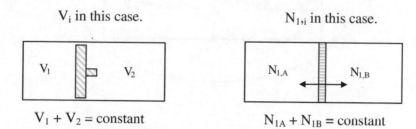

V_i in this case.	$N_{1,i}$ in this case.

$V_1 + V_2 = $ constant $N_{1A} + N_{1B} = $ constant

Mathematically, the **Entropy Maximum Principle** can be expressed as follows:

$$S = S(U, V, N_1, \dots N_r)$$

For a given value of the total internal energy, i.e., $U_T = $ constant,

$$\left(\frac{\partial S}{\partial X} \right)_{U_T} = 0 \quad \text{and} \quad \left(\frac{\partial^2 S}{\partial X^2} \right)_{U_T} < 0$$

where $X = V, N_1, \dots N_r$.

It can be shown that the entropy maximum principle is equivalent to the **Energy Minimum Principle**: **For a simple, isolated system, the equilibrium value of any unconstrained internal parameters will minimize the energy under the given value of the total entropy**.

Mathematically, the **Energy Minimum Principle** can be expressed as follows:

$$U = U(S, V, N_1, \dots N_r)$$

For a given value of the total entropy, i.e., $S_T = $ constant,

$$\left(\frac{\partial U}{\partial X}\right)_{S_T} = 0 \quad \text{and} \quad \left(\frac{\partial^2 U}{\partial X^2}\right)_{S_T} > 0$$

where $X = V, N_1, \ldots N_r$.

Here a simple example will be used to demonstrate the equivalence of the energy minimum principle and the entropy maximum principle.

For an isolated system consisting of two subsystems in thermal contact, we have proved by using entropy maximum principle that the equilibrium condition is $T_1 = T_2$. Now, let's use the energy minimum principle to find the equilibrium condition.

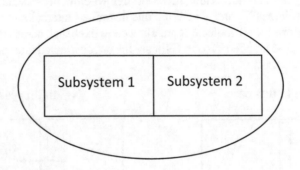

Recall

$$U = U(S, V, N_1, \ldots, N_r) = TS - PV + \sum \mu_i N_i$$

$$dU = TdS - PdV + \sum \mu_i dN_i$$

and the constraints for this system are:

$$V_1 = \text{constant} \quad V_2 = \text{constant}$$
$$N_{1k} = \text{constant} \quad N_{2k} = \text{constant} \quad (k = 1, 2, \ldots r)$$

and

$$S_1 + S_2 = \text{constant}.$$

The constraint conditions lead to

$$dV_i = 0, \quad \text{and} \quad dN_k = 0$$

Therefore,

$$dU = TdS$$

$$dU_{total} = dU_1 + dU_2 = T_1\, dS_1 + T_2\, dS_2$$

Using the constraint

$$S_1 + S_2 = \text{constant},$$

$$dS_2 = -dS_1$$

we have

$$dU_{Total} = T_1 dS_1 + T_2 dS_2 = (T_1 - T_2) dS_1$$

According to the energy minimum energy principle, at equilibrium state,

$$dU_{Total} = 0,$$

it follows that

$$T_1 = T_2.$$

This is the same conclusion as the equilibrium condition derived by using the entropy maximum principle. Therefore, **one can use either the entropy maximum principle or the energy minimum principle to find the equilibrium conditions of a thermodynamics system.**

1.7 Legendre Transformations and Thermodynamic Potentials

Legendre Transformation

Previously, we have shown that we can find the equilibrium conditions by using the entropy maximum principle or energy minimum principle and by using one of the following fundamental equations.

$$S = S(U, V, N_1, \ldots N_r) \qquad dS = \frac{1}{T}dU + \frac{P}{T}dV - \Sigma \frac{\mu_i}{T}dN_i$$
$$U = U(S, V, N_1, \ldots N_r) \qquad dU = TdS - PdV + \Sigma \mu_i\, dN_i$$

These are the most basic approaches in thermodynamic analysis. In practice, however, we will find that using these extensive variables (S, U, V, N_i) is not always the most convenient way to model and to solve problems. We often need to use different combinations of some intensive variables and some extensive variables.

For example, for an isothermal process, it will be more convenient that if we can use the intensive variable, temperature T, instead of the entropy S, with other extensive variables to model such a process. Because T is a constant in this case and hence such a model with this group of variables will reduce the number of independent variables by one. In order to do so, a proper mathematical transformation of variables is required. In other words, we need to transfer a function with one group of variables into another function with a different group of variables.

$$(X, Y) \text{ domain} \quad \xrightarrow{\textbf{Transformation}} \quad (\zeta, \eta) \text{ domain}$$

Let us consider a function Y given by:

$$Y = Y(X_1, \ldots, X_n)$$

and let

$$P_k = \frac{\partial Y}{\partial X_k}$$

It follows that

$$dY = \sum_{k=1}^{n} P_k dX_k$$

Let us define a new function:

$$\psi = Y - \sum_{k=1}^{n} P_k X_k$$

Differentiate the above equation,

$$d\psi = dY - \Sigma P_k dX_k - \Sigma X_k dP_k$$

Recall $dY = \Sigma P_k dX_k$.
The above equation is reduced to

$$d\psi = -\sum_{k=1}^{n} X_k dP_k$$

From this equation, we can conclude that the new function ψ is a function of variables $P_1, P_2, \ldots P_n$. That is,

$$\psi = \psi(P_1, P_2, \ldots, P_n)$$

As seen from the above, we have transferred the function

$$Y(X_1, \ldots, X_n)$$

into a new function

$$\psi(P_1, \ldots, P_n)$$

with a new set of variables. This transform is referred to as the **Legendre Transformation**.

Legendre transform can also be made only for some of the variables. For example, we want to keep a part of the original variables, $X_1, X_2, \ldots\ldots X_k$, and transfer X_{k+1}, $X_{k+2}, \ldots\ldots X_n$ into $P_{k+1}, P_{k+2}, \ldots\ldots P_n$. This can be realized in the following way. Define a new function as:

$$\psi = \psi(X_1, X_2, \ldots, X_k, P_{k+1}, \ldots, P_n) = Y - \sum_{k+1}^{n} P_i X_i$$

where

$$P_i = \frac{\partial Y}{\partial X_i}$$

This is called the **partial Legendre transformation**. Now let us apply the partial Legendre transformation to the internal energy function U to derive some useful thermodynamic functions.

Thermodynamic Potentials

Helmholtz Free Energy/Potential

Let us start with the energy form of the fundamental equation as given below

$$U = U(S, V, N_1, \ldots, N_r) = TS - PV + \sum \mu_i N_i$$

$$dU = TdS - PdV + \sum \mu_i dN_i$$

In the above fundamental equation, entropy S is not a directly measurable parameter and is often not convenient to use as an extensive variable, therefore we want to replace S by the corresponding intensive variable, temperature T, while keeping the rest of variables. It should be noted that here the internal energy $U = U(S, V, N_1 \cdots N_r)$ corresponds to the original function $Y = Y(X_1, \cdots X_n)$ in the Legendre transformation. According to

$$\psi = Y - \sum_{k+1}^{n} P_i X_i \qquad \text{and} \qquad P_i = \frac{\partial Y}{\partial X_i}$$

Performing the partial Legendre transformation to the internal energy U by defining a new function F as follows:

$$F = U - \left(\frac{\partial U}{\partial S}\right) S = U - TS$$

This function is called Helmholtz free energy or Helmholtz potential. Recall

$$U = TS - PV + \sum_i \mu_i N_i$$

we have

$$F = -PV + \sum \mu_i N_i$$

Differentiating $F = U - TS$ yields

$$dF = dU - d(TS)$$
$$= TdS - PdV + \sum \mu_i dN_i - TdS - SdT$$
$$= -SdT - PdV + \sum \mu_i dN_i$$

The above differential equation clearly indicates that

$$F = F(T, V, \{N_i\})$$

That is, Helmholtz free energy is a function of temperature T, volume V and mole numbers N_i. Comparing it with

$$U = U(S, V, \{N_i\})$$

we see that T in the new function F replace S in the original function U. For a constant temperature process, i.e., $T = $ constant,

$$F = F(T = c, V, \{N_i\}),$$

$$dF = -PdV + \sum \mu_i dN_i$$

For a process with both constant T and constant V,

$$dF = \sum \mu_i \, dN_i$$

Enthalpy

In some applications, we want to replace volume V in the internal energy function by the corresponding intensive variable, pressure P. Applying the Legendre transformation

$$\psi = Y - \sum P_i \, X_i$$

to the internal energy function

$$U = U(S, V, \{N_i\}) \quad \text{and} \quad U = TS - PV + \sum_i \mu_i \, N_i,$$

we define a new function.

$$H = U - \left(\frac{\partial U}{\partial V}\right)V$$

$$H = U + PV = TS + \sum \mu_i \, N_i$$

The function H is called the **enthalpy**. Differentiating H function gives

$$dH = dU + PdV + VdP = TdS + VdP + \sum \mu_i \, dN_i$$

It follows that

$$H = H(S, P, \{N_i\})$$

Comparing it with

$$U = U(S, V, \{N_1\})$$

we see that P in the H function replaces V in the U function.

For a constant pressure process, i.e., P = constant,

$$dH = TdS + \sum \mu_i \, dN_i$$

Gibbs Free Energy/Potential

Now we want to replace both S by T and V by P. Let us perform the Legendre transformation

$$\psi = Y - \Sigma P_i X_i$$

to $U = (S, V, N_1, N_2, ..., N_r)$ by defining a new function

$$G = U - \left(\frac{\partial U}{\partial S}\right)S - \left(\frac{\partial U}{\partial V}\right)V = U - TS + PV = \sum \mu_i N_i$$

The function G is called the Gibbs free energy or Gibbs potential. For a single component system where $r = 1$,

$$G = \mu N,$$

and

$$\mu = G/N = g.$$

That is why the chemical potential sometimes is also called the specific Gibbs free energy (i.e., Gibbs free energy per unit mole). The differential form of Gibbs free energy is given below.

$$dG = dU - TdS - SdT + PdV + VdP = -SdT + VdP + \sum \mu_i dN_i$$

This differential equation suggests that

$$G = G(T, P, \{N_i\})$$

that is, the Gibbs free energy is a function of temperature T, pressure P and mole numbers N_i.

Comparing it with

$$U = U(S, V, \{N_1\})$$

we see that T and P in the G function replace the S and V in the U function. For a process with constant T and constant P, i.e., $T = $ const. and $P = $ const., we have

$$dG = \sum \mu_i dN_i$$

For a process with $N_i = $ constant, we have

$$dG = -SdT + VdP$$

Grand Canonical Free Energy/Potential

For many applications, the system is subject to a constant temperature and a constant chemical potential. Therefore, we want to define a new thermodynamic potential— grand canonical potential (or grand canonical free energy) that includes T and μ in the variables.

Recall

$$U = U(S, V, N_1, \ldots, N_r) = TS - PV + \sum \mu_i N_i$$

$$dU = TdS - PdV + \sum \mu_i dN_i$$

Apply the partial Legendre transformation to the internal energy function U,

$$\psi = Y - \sum_{k+1}^{n} P_i X_i \qquad \text{and} \qquad P_i = \frac{\partial Y}{\partial X_i}$$

the grand canonical free energy function is defined as the following:

$$\Omega = U - TS - \sum \mu_i N_i = -PV$$

$$d\Omega = dU - d(TS) - d(\sum \mu_i N_i)$$

$$= -SdT - PdV - \sum N_i \, d\mu_i$$

As seen from the above equation, the grand canonical free energy is a function of temperature T, volume V and chemical potentials μ_i,

$$\Omega = \Omega(T, V, \{\mu_i\})$$

Comparing it with

$$U = U(S, V, \{N_i\})$$

we see that T and μ_i in the Ω function replace S and N_i in the U function. For a process with constant T and μ_i (i.e., thermal and chemical equilibrium), the change in the grand canonical free energy is the mechanical work,

$$d\Omega = -PdV = dW.$$

The specific grand canonical free energy per unit volume is defined as

$$\omega = \frac{\Omega}{V} = -P.$$

1.8 Minimum Principles of Thermodynamic Potentials/Free Energies

Helmholtz Free Energy Minimum Principle

Consider a system in contact with a thermal reservoir as shown in the following figure. The combined system (i.e., the system plus the reservoir) is an isolated system. The properties of the reservoir are indicated by a superscript r. From the energy minimum principle, we know that, at a given value of the total entropy, $S_T = S + S^r = $ constant, the equilibrium state requires:

$$dU_{total} = d(U + U^T) = 0$$

and

$$d^2 U_{total} = d^2(U + U^T) > 0$$

where U is the internal energy of the system, and U^T is the internal energy of the thermal reservoir.

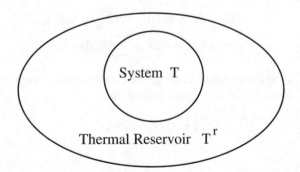

A thermal reservoir is an idealized system that contains infinite amount of thermal energy (heat) and has a constant temperature. Its sole function is to exchange heat with any interacting system so that, at equilibrium, the interacting system will have the same temperature as the thermal reservoir. The fundamental equation of a thermal reservoir is given by:

$$U^T = U^T(S^r)$$

and

$$dU^T = T^r dS^r.$$

The total internal energy of this combined isolated system is:

$$U_{total} = U + U^T$$

It follows that

$$dU_{total} = d\left(U + U^T\right) = dU + dU^T = dU + T^r dS^r$$

Because

$$S_T = S + S^r = \text{constant,}$$

$$dS^r = -dS$$

It follows that

$$dU_{total} = d\left(U + U^T\right) = dU + dU^T = dU + T^r dS^r = dU - T^r dS$$

Realizing that the system is in contact with a thermal reservoir and reaches an equilibrium state, $T^r = T = \text{constant}$, we have

$$dU_{total} = d\left(U + U^T\right) = dU + dU^T = dU + T^r dS^r = dU - T^r dS$$
$$= dU - TdS = d(U - TS) = dF = 0$$

Similarly, we can show

$$d^2U^T = d\left(dU^T\right) = d(T^r dS^r) = d(-T^r dS) = -T^r d^2 S = -d^2(T^r S) = -d^2(TS)$$

$$d^2U_{total} = d^2\left(U + U^T\right) = d^2U + d^2U^T = d^2U - d^2(TS) = d^2(U - TS) = d^2F > 0$$

As shown above, we have proved that the Helmholtz potential F will be minimum at a stable equilibrium state, if the temperature is constant. The **Helmholtz free energy minimum principle** can be stated as follows:

The equilibrium values of any unconstrained internal parameters of a system in thermal contact with a thermal reservoir minimize the Helmholtz free energy when T = T^r = constant.

Because $T = \text{constant}$, the unconstrained internal parameters are the volume, V, and the mole numbers N_i in $F = F(\text{T}, \text{V}, \{N_i\})$ function.

Gibbs Free Energy Minimum Principle

Consider a system in contact with a thermal reservoir and a pressure reservoir as shown in the figure below. A pressure reservoir is an idealized system that has a constant pressure. Its sole function is to exchange volume with any interacting system so that, at equilibrium, the interacting system will have the same pressure as the pressure reservoir.

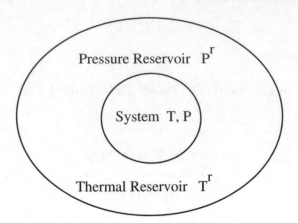

The fundamental equation for a thermal reservoir is:

$$U^T = U(S^r) \qquad dU^T = T^T dS^r$$

The fundamental equation for a pressure reservoir is:

$$U^P = U(V^r), \quad dU^P = -P^r dV^r$$

The total internal energy of the combined system is:

$$U_{total} = U + U^{T^r} + U^{P^r}$$

At equilibrium state, the energy minimum principle requires

$$dU_{total} = d\left(U + U^{T^r} + U^{P^r}\right) = 0$$

and

$$d^2 U_{total} = d^2\left(U + U^{T^r} + U^{P^r}\right) > 0$$

The constraints of the combined system (the system and the reservoirs) are

$$V + V^r = \text{constant}, \qquad S + S^r = \text{constant}.$$

or

$$dV = -dV^r, \qquad dS = -dS^r$$

Because, at an equilibrium state,

$$T^r = \text{constant} = T \qquad P^r = \text{constant} = P$$

It follows

$$\begin{aligned}
d(U + U^{T^r} + U^{P^r}) &= dU + T^r dS^r - P^r dV^r \\
&= dU - T^r dS + P^r dV \\
&= d(U - TS + PV) \\
&= dG
\end{aligned}$$

According to the energy minimum principle:

$$d(U + U^{T^r} + U^{P^r}) = 0$$

we have

$$dG = 0$$

Furthermore, we can show

$$\begin{aligned}
d^2 G &= d^2(U - TS + PV) \\
&= d(dU - TdS + PdV) \\
&= d(dU + TdS^r - PdV^r) \\
&= d^2(U + T^r S^r - P^r V^r) \\
&= d^2(U + U^{T^r} + U^{P^r}) > 0.
\end{aligned}$$

From the above equations, we have proved that the Gibbs free energy is minimum at equilibrium if both temperature and pressure are kept constant. The **Gibbs free energy minimum principle** can be stated as follows:

The equilibrium values of any unconstrained internal parameters in a system in contact with a thermal reservoir and a pressure reservoir minimize the Gibbs free energy when $T = T^r = $ constant and $P = P^r = $ constant.

Because T and P must be constant as the system is in contact with a thermal reservoir and a pressure reservoir, the unconstrained internal parameters are the mole numbers N_i in $G = G(T, P, \{N_i\})$ function.

Grand Canonical Free Energy Minimum Principle

Consider a system in contact with a thermal reservoir and a mass reservoir as illustrated in the following figure. The mass reservoir is an idealized system that has infinite amount of molecules and constant chemical potentials for each type of molecules. Its sole function is to exchange molecules with the interacting system so that, at equilibrium, the interacting system will have the same values of chemical potentials as that of the mass reservoir. For the thermal reservoir and the mass reservoir, the fundamental equations are:

$$dU^T = T^r \, dS^r$$

$$dU^M = \sum \mu_i^r \, dN_i^r$$

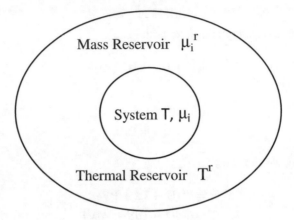

The constraints for the combined system (i.e., the system and the reservoirs).

$$S + S^r = \text{constant}$$
$$N_i + N_i^r = \text{constant} \quad (i = 1, 2, \dots r)$$

At equilibrium states, we know

$$T = T^r = \text{constant}$$
$$\mu_i = \mu_i^r = \text{constant}$$

Thus, the total internal energy of the combined system is:

$$\begin{aligned}
d(U + U^{T^r} + U^{M^r}) &= dU + T^r \, dS^r + \sum \mu_i^r \, dN_i^r \\
&= dU - T^r \, dS - \sum \mu_i^r \, dN_i \\
&= d(U - T^r \, S - \sum \mu_i^r \, N_i) \\
&= d(U - TS - \sum \mu_i \, N_i) \\
&= d\Omega
\end{aligned}$$

Because the combined system is an isolated system, the energy minimum principle requires

$$d(U + U^{T^r} + U^{M^r}) = d\Omega = 0$$

Similarly, we can show

$$d^2\Omega = d^2(U + U^{T^r} + U^{M^r}) > 0$$

Thus, we have proved that Ω free energy/potential is minimum at equilibrium states if the temperature and the chemical potentials are constant. The **Grand Canonical Free Energy Minimum Principle** may be stated as follows:

The equilibrium values of any unconstrained internal parameters in a system in contact with a thermal reservoir and a mass reservoir minimize the Grand Canonical free energy when $T = T^r = $ constant and $\mu_i = \mu_i^r = $ constant.

The unconstrained internal parameter is V in $\Omega = \Omega(T, V, \{\mu_i\})$ function.

In summary, when you need to find thermodynamic equilibrium conditions by using a thermodynamic potential,

- Use Helmholtz potential minimum principle if the system is in contact with a thermal reservoir, i.e., temperature is constant.
- Use Gibbs free energy minimum principle if the system is in contact with a thermal reservoir and a pressure reservoir, i.e., $T = $ constant and $P = $ constant.
- Use Grand Canonical free energy minimum principle if the system is in contact with a thermal reservoir and a mass reservoir, i.e., $T = $ constant and $\mu_i = $ constant.

1.9 Applications of Minimum Principle of Thermodynamic Potentials

Helmholtz Potential Minimum Principle

Consider a composite system consisting of two subsystems separated by a partition, as illustrated in the figure below. In subsystem A (left side), there is a dilute aqueous solution consisting of water and a salt (e.g., NaC1). There is only water in subsystem B (right side). However, the water presents in two phases (i.e., liquid and vapor) in subsystem B. Assume that the total volume and the total mass of the system are constant. The system is surrounded by a thermal reservoir.

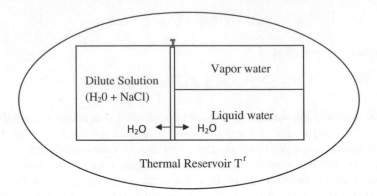

The constraints for the combined system (the system and the reservoir) are:

$$U + U^R = \text{constant}$$
$$N_i = \text{constant}, \qquad i = 1, 2, \ldots r$$
$$V_{Total} = \text{constant}$$

The internal constraints are (1) the subsystems are separated by a rigid partition so that the volumes of the subsystems are constant. (2) The partition is semi-permeable, and only water molecules can pass through the partition.

$$N_{AW} + N_{BW}^L + N_{BW}^V = \text{constant}$$
$$N_{solute} = \text{constant},$$
$$V_A = \text{constant}$$
$$V_B = V_{BL} + V_{BV} = \text{constant}$$

where the subscripts A and B stand for the subsystem A and subsystem B; the subscript W stands for water; and the subscript or superscripts L and V stand for liquid and vapor, respectively.

We would like to find (1) what thermodynamic function should be used as the thermodynamic potential to model this system, and (2) the equilibrium conditions for this system.

Solution: To determine the equilibrium conditions that the intensive properties must satisfy by using the minimum principle of thermodynamic potentials, the first question is "Which thermodynamic potential should we use to model this system"? *The answer is determined by the constraints between the system and the reservoir.* For the system described above, since it is in contact with a thermal reservoir, the temperature is constant. Under this condition, Helmholtz free energy or Helmholtz potential will be minimum at equilibrium state, as demonstrated in the previous section. Therefore, we will use Helmholtz potential as the thermodynamic potential to model this system and find the equilibrium conditions.

For this composite system, the total Helmholtz potential F is:

$$F = F_A + F_B^L + F_B^V$$
$$dF = dF_A + dF_B^L + dF_B^L$$

And individually,

$$dF_A = - S_A \, dT - P_A \, dV_A + \mu_{AW} \, dN_{AW} + \mu_S \, dN_S$$
$$dF_B^L = - S_B^L \, dT - P_B^L \, dV_B^L + \mu_{BW}^L \, dN_{BW}^L$$
$$dF_B^V = - S_B^V \, dT - P_B^V \, dV_B^V + \mu_{BW}^V \, dN_{BW}^V$$

where the subscripts and superscripts W, S, L, V represent water, salt, liquid water and vapor water, respectively. According to the given constraints, we have.

$$T = \text{constant}, \quad V_A = \text{constant}, \quad \text{and} \quad N_S = \text{constant}.$$

This leads to:

$$dF = \mu_{AW} dN_{AW} - P_B^L \, dV_B^L + \mu_{BW}^L \, dN_{BW}^L - P_B^V \, dV_B^V + \mu_{BW}^V \, dN_{BW}^V$$

Using the constraints:

$$N_W = N_W^A + N_{BW}^L + N_{BW}^V = \text{constant}, \quad \text{i.e.,} \quad dN_W^A = - dN_{BW}^L - dN_{BW}^V$$
$$V_B^L + V_B^V = \text{constant}, \quad \text{i.e.,} \quad d V_B^V = -d V_B^L$$

We have

$$dF = (P_B^V - P_B^L) \, dV_B^L + (\mu_{BW}^L - \mu_{AW}) \, dN_{BW}^L + (\mu_{BW}^V - \mu_{AW}) \, dN_{BW}^V$$

where dV_B^L, dN_W^{BL} and dN_W^{BV} are unconstrained variables.

Because total Helmholtz potential F of this system must be minimum at equilibrium states, i.e.,

$$dF = 0,$$

it follows that the equilibrium conditions are

$$P_B^L = P_B^V$$
$$\mu_{AW} = \mu_{BW}^L = \mu_{BW}^V$$

Gibbs Potential Minimum Principle

Consider a liquid–gas two-phase system enclosed in a piston-cylinder device in contact with a thermal reservoir and a pressure reservoir, as shown below. The piston is free to move without any friction. Therefore, the pressure inside the piston-cylinder devise is balanced with the pressure of the pressure reservoir. The liquid phase contains two components (1 and 2) and the gas phase has only the component 2. We want to find equilibrium conditions. For such a system, since $T = $ constant, $P = $ constant, obviously, using Gibbs free energy or Gibbs potential to model the composite system is the right choice.

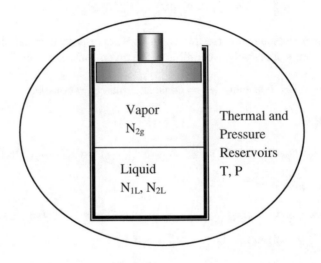

$$G = G_g + G_L,$$
$$G_g = \mu_2 \, N_{2g},$$
$$G_L = \mu_{1L} \, N_{1L} + \mu_{2L} \, N_{2L}$$

Because the differential form of the Gibbs free energy is given by:

$$dG = -SdT + VdP + \sum \mu_i \, dN_i$$

When $dT = dP = 0$,

$$dG = \sum \mu_i dN_i$$

For this given system, the total Gibbs free energy or total Gibbs potential is

$$G = G_L + G_g$$

Therefore,

$$dG = dG_L + dG_g = \mu_{1L}dN_{1L} + \mu_{2L}dN_{2L} + \mu_{2g}dN_{2g}$$

Use the constraints:

$$N_{2g} + N_{2L} = \text{constant}, \qquad N_{1L} = \text{constant}$$

$$dG = \mu_{2g}\,dN_{2g} + \mu_{2L}\,dN_{2L} = (\mu_{2G} - \mu_{2L})\,dN_{2g}$$

At equilibrium, the total Gibbs potential of this system must be minimum, i.e.,

$$dG = 0$$

it follows that the equilibrium condition is

$$\mu_{2g} = \mu_{2L}$$

In general, we can show that the condition

$$\mu_i^{\alpha} = \mu_i^{\beta} = \ldots = \mu_i^{\gamma} \quad (i = 1, 2, \ldots r)$$

is true for a multi-component (i.e., r components) and multi-phase ($\alpha, \beta, \ldots \gamma$) system in equilibrium states, as long as the molecules exchange is allowed between two neighboring phases.

1.10 Maxwell Relations

The mathematical properties, especially the continuity condition, of the fundamental equations require the equalities of the mixed partial derivatives of fundamental equations (including various Legendre transformations of the fundamental equations, i.e., thermodynamic potentials).

Let

$$Z = Z(\{X_i\})$$

be the general form of the fundamental equation U and the thermodynamic potential functions, the equalities of the mixed partial derivatives are given in the following form:

$$\frac{\partial^2 Z}{\partial X_i \partial X_j} = \frac{\partial^2 Z}{\partial X_j \partial X_i}$$

where Z may be U, H, F, G, and Ω. These equalities are called the "**Maxwell Relations**". For example, for a single-component system with constant mass, from

$$U = U(S, V, N = \text{constant})$$

$$dU = \left(\frac{\partial U}{\partial S}\right)_V dS + \left(\frac{\partial U}{\partial V}\right)_S dV = TdS - PdV$$

the Maxwell relation requires

$$\left(\frac{\partial^2 U}{\partial S \partial V}\right) = \left(\frac{\partial^2 U}{\partial V \partial S}\right)$$

That is,

$$\left(\frac{\partial T}{\partial V}\right)_S = -\left(\frac{\partial P}{\partial S}\right)_V$$

Similarly, from

$$dH = TdS + VdP$$
$$dF = -SdT - PdV$$
$$dG = -SdT + VdP$$

the Maxwell relations require

$$\left(\frac{\partial T}{\partial P}\right)_S = \left(\frac{\partial V}{\partial S}\right)_P$$
$$\left(\frac{\partial P}{\partial T}\right)_V = \left(\frac{\partial S}{\partial V}\right)_T$$
$$\left(\frac{\partial V}{\partial T}\right)_P = -\left(\frac{\partial S}{\partial P}\right)_T$$

One of the important applications of Maxwell Relations is to express the derivatives in thermodynamic relationships in terms of measurable parameters. Let's first introduce some thermodynamic parameters.

Coefficient of thermal expansion

$$\alpha = \frac{1}{V}\left(\frac{\partial V}{\partial T}\right)_P$$

Isothermal compressibility

$$\kappa_T = -\frac{1}{V}\left(\frac{\partial V}{\partial P}\right)_T$$

Adiabatic compressibility

$$\kappa_s = -\frac{1}{V}\left(\frac{\partial V}{\partial P}\right)_S$$

Molar heat capacities

$$C_P = T\left(\frac{\partial s}{\partial T}\right)_P = \frac{T}{N}\left(\frac{\partial S}{\partial T}\right)_P$$

$$C_V = T\left(\frac{\partial s}{\partial T}\right)_V = \frac{T}{N}\left(\frac{\partial S}{\partial T}\right)_V$$

The following are some useful mathematical relations:

$$\left(\frac{\partial X}{\partial Y}\right)_Z = \frac{1}{\left(\frac{\partial Y}{\partial X}\right)_Z}$$

$$\left(\frac{\partial X}{\partial Y}\right)_Z = \frac{\left(\frac{\partial X}{\partial w}\right)_Z}{\left(\frac{\partial Y}{\partial w}\right)_Z}$$

$$\left(\frac{\partial X}{\partial Y}\right)_Z = -\frac{\left(\frac{\partial Z}{\partial Y}\right)_X}{\left(\frac{\partial Z}{\partial X}\right)_Y} \quad \text{or} \quad \left(\frac{\partial X}{\partial Y}\right)_Z\left(\frac{\partial Y}{\partial Z}\right)_X\left(\frac{\partial Z}{\partial X}\right)_Y = -1$$

The last relationship is called the **chain rule**.

Example 1: Prove the following equations for a simple compressible system:

(a) $\left[\frac{\partial C_V}{\partial v}\right]_T = T\left[\frac{\partial}{\partial T}\left(\frac{\alpha}{\kappa_T}\right)\right]_V$

(b) $\left[\frac{\partial C_P}{\partial P}\right]_T = -T\left[\frac{\partial}{\partial T}(\alpha v)\right]_P$

(c) $C_P - C_V = \frac{\alpha^2 vT}{\kappa_T}$

where α and κ_T are the coefficients of thermal expansion and the isothermal compressibility respectively.

(a)   Solution: Recall

$$\alpha = \frac{1}{v}\left(\frac{\partial v}{\partial T}\right)_P$$

$$\kappa_T = -\frac{1}{v}\left(\frac{\partial v}{\partial P}\right)_T$$

$$C_V = T\left(\frac{\partial s}{\partial T}\right)_V$$

Using the chain rule gives

$$\left(\frac{\partial P}{\partial T}\right)_v \left(\frac{\partial T}{\partial v}\right)_P \left(\frac{\partial v}{\partial P}\right)_T = -1$$

we have

$$\frac{\alpha}{k_T} = \frac{\frac{1}{V}\left(\frac{\partial V}{\partial T}\right)_P}{-\frac{1}{V}\left(\frac{\partial V}{\partial P}\right)_T} = \frac{\left(\frac{\partial V}{\partial T}\right)_P}{\left(\frac{\partial P}{\partial T}\right)_V \left(\frac{\partial T}{\partial V}\right)_P} = \left(\frac{\partial V}{\partial T}\right)_P \left(\frac{\partial P}{\partial T}\right)_V \left(\frac{\partial T}{\partial V}\right)_P = \left(\frac{\partial P}{\partial T}\right)_V$$

Inspecting the right-hand side of the above equation, we see that T and v are variables. This leads us to think about the specific Helmholtz potential, $f(T, v)$. From

$$df = -sdT - Pdv$$

Maxwell Relation in the *df* function is

$$\left(\frac{\partial P}{\partial T}\right)_v = \left(\frac{\partial s}{\partial v}\right)_T$$

That leads to:

$$\frac{\alpha}{\kappa_T} = \left(\frac{\partial s}{\partial v}\right)_T$$

Then,

$$\left[\frac{\partial}{\partial T}\left(\frac{\alpha}{\kappa_T}\right)\right]_v = \left[\frac{\partial}{\partial T}\left(\frac{\partial s}{\partial v}\right)_T\right]_v = \left[\frac{\partial}{\partial v}\left(\frac{\partial s}{\partial T}\right)_v\right]_T = \left[\frac{\partial}{\partial v}\left(\frac{C_v}{T}\right)\right]_T = \frac{1}{T}\left(\frac{\partial C_v}{\partial v}\right)_T$$

That is

$$T\left[\frac{\partial}{\partial T}\left(\frac{\alpha}{\kappa_T}\right)\right]_v = \left(\frac{\partial C_v}{\partial v}\right)_T$$

(b)   Solution: Because

$$\alpha v = \left(\frac{\partial v}{\partial T}\right)_P$$

inspecting the right-hand side of the above equation, we see that T and P are variables. This leads us to think of the specific Gibbs potential, $g(T, P)$. From

$$dg = -sdT + vdP$$

we have the following Maxwell relation:

$$\left(\frac{\partial v}{\partial T}\right)_P = -\left(\frac{\partial s}{\partial P}\right)_T$$

Thus

$$\left[\frac{\partial}{\partial T}(\alpha v)\right]_P = -\left[\frac{\partial}{\partial T}\left(\frac{\partial s}{\partial P}\right)_T\right]_P = -\left[\frac{\partial}{\partial P}\left(\frac{\partial s}{\partial T}\right)_P\right]_T$$

$$= -\left[\frac{\partial}{\partial P}\left(\frac{C_P}{T}\right)\right]_T = -\frac{1}{T}\left(\frac{\partial C_P}{\partial P}\right)_T$$

Finally,

$$-T\left[\frac{\partial}{\partial T}(\alpha v)\right]_P = \left(\frac{\partial C_P}{\partial P}\right)_T$$

(c)   Solution: Because

$$C_p - C_v = T\left[\left(\frac{\partial s}{\partial T}\right)_P - \left(\frac{\partial s}{\partial T}\right)_v\right]$$

The differential of the specific entropy may be written as

$$ds = \left(\frac{\partial s}{\partial T}\right)_v dT + \left(\frac{\partial s}{\partial v}\right)_T dv$$

$$\left(\frac{\partial s}{\partial T}\right)_P = \left(\frac{\partial s}{\partial T}\right)_v + \left(\frac{\partial s}{\partial v}\right)_T\left(\frac{\partial v}{\partial T}\right)_P$$

Therefore,

$$C_P - C_v = T\left(\frac{\partial s}{\partial v}\right)_T\left(\frac{\partial v}{\partial T}\right)_P$$

From

$$df = -sdT - Pdv$$

we have

$$\left(\frac{\partial s}{\partial v}\right)_T = \left(\frac{\partial P}{\partial T}\right)_v$$

It follows

$$C_P - C_v = T\left(\frac{\partial P}{\partial T}\right)_v \left(\frac{\partial v}{\partial T}\right)_P$$

Using the partial differential identity:

$$\left(\frac{\partial P}{\partial T}\right)_v = -\frac{\left(\frac{\partial v}{\partial T}\right)_P}{\left(\frac{\partial v}{\partial P}\right)_T}$$

we finally have

$$C_P - C_v = -T\frac{\left(\frac{\partial v}{\partial T}\right)_P \left(\frac{\partial v}{\partial T}\right)_P}{\left(\frac{\partial v}{\partial P}\right)_T} = \frac{Tv\,\alpha^2}{\kappa_T}$$

Example 2: Prove the following equations for a simple compressible system:

$$du = C_v\,dT + \left[\frac{\alpha T}{\kappa_T} - P\right]dv$$

Solution: Assuming $u = u\,(T, v)$, we have

$$du = \left(\frac{\partial u}{\partial T}\right)_v dT + \left(\frac{\partial u}{\partial v}\right)_T dv$$

Using $du = Tds - Pdv$.
We have

$$\left(\frac{\partial u}{\partial T}\right)_v = T\left(\frac{\partial s}{\partial T}\right)_v - P\left(\frac{\partial v}{\partial T}\right)_v = T\left(\frac{\partial s}{\partial T}\right)_v - 0 = C_v$$

$$\left(\frac{\partial u}{\partial v}\right)_T = \underbrace{T\left(\frac{\partial s}{\partial v}\right)_T} - P = T\left(\frac{\alpha}{\kappa_T}\right) - P$$

$$\frac{\alpha}{\kappa_T} = \left(\frac{\partial s}{\partial v}\right)_T$$

Combining the above two equations with the 1st equation, we will have

$$du = C_v\,dT + \left(\frac{T\alpha}{\kappa_T} - P\right)dv$$

Example 3: Show $\left[\frac{\partial P}{\partial V}\right]_s = \frac{-C_P}{C_v \kappa_T V}$.

Solution: For a simple system with a fixed mass, from

$$dF = -SdT - PdV$$

the Maxwell relations is

$$\left(\frac{\partial S}{\partial V}\right)_T = \left(\frac{\partial P}{\partial T}\right)_V$$

From

$$dG = -SdT + VdP$$

the Maxwell relations is

$$-\left(\frac{\partial S}{\partial P}\right)_T = \left(\frac{\partial V}{\partial T}\right)_P$$

Recall

$$\alpha = \frac{1}{V}\left(\frac{\partial V}{\partial T}\right)_P \qquad \kappa_T = -\frac{1}{V}\left(\frac{\partial V}{\partial P}\right)_T$$

We will need to use the above relationships in the following derivation. Because

$$\left(\frac{\partial P}{\partial V}\right)_{s,} = \left(\frac{\partial P}{\partial T}\right)_{s,}\left(\frac{\partial T}{\partial V}\right)_{s,}$$

$$\left(\frac{\partial T}{\partial V}\right)_s = -\frac{\left(\frac{\partial S}{\partial V}\right)_T}{\left(\frac{\partial S}{\partial T}\right)_V} = -\frac{\left(\frac{\partial P}{\partial T}\right)_V}{\frac{C_v}{T}} = \frac{\left(\frac{\partial P}{\partial V}\right)_T}{\frac{C_v}{T}\left(\frac{\partial T}{\partial V}\right)_P} = -\frac{T\alpha}{C_v \kappa_T}$$

$$\left(\frac{\partial P}{\partial T}\right)_s = -\frac{\left(\frac{\partial S}{\partial T}\right)_P}{\left(\frac{\partial S}{\partial P}\right)_T} = -\frac{\frac{C_p}{T}}{-\left(\frac{\partial V}{\partial T}\right)_P} = \frac{C_P}{TV\alpha}$$

Finally,

$$\left(\frac{\partial P}{\partial V}\right)_s = \left[\frac{C_P}{TV\alpha}\right]\left[\frac{-T\alpha}{C_v \kappa_T}\right] = -\frac{C_P}{VC_v \kappa_T}$$

1.11   Thermodynamic Characteristics of Dielectric Media

Application of Maxwell Relations

Dielectric materials include plastics, organic liquids, water (including aqueous electrolyte solutions) and gases. A dielectric material is poor electrical conductor and can be polarized by an applied electric field. Molecules of many dielectric materials are permanently polarized due to their asymmetrical molecular structure. A simple example is HCl. The permanent dipoles will be aligned with the externally applied electric field (dipole re-orientation). Dielectric materials containing symmetrical molecules or atoms can also become polarized when exposed to an electrical field, resulting from the relative displacement of orbital electrons, such as the case of He. That is, when a dielectric is under an electric field, electric charges will be shifted a little from their equilibrium positions. Positive charges move towards the negative electrode and negative charges move towards the positive electrode (local charge migration). This is called the polarization. Therefore, in the presence of an external electrical field, molecules of all dielectric materials have a dipole (i.e., are polarized). When we consider a dielectric material in an applied electric field, we must consider the electric work due to the interaction of the molecular dipole with the applied electric field.

The differential form of the fundamental equation for a dielectric medium is:

$$dU = TdS - P_0 \, dV + \frac{V}{4\pi} \vec{E} \cdot d\vec{D}$$

where $\vec{E}$ is the applied electrical field strength, $\vec{D}$ is the dielectric displacement, P_0 is the pressure. It should be noted that

$$\vec{D} = \vec{E} + 4\pi \vec{P},$$

where $\vec{P}$ is the polarization density per unit volume, and these properties ($\vec{D}$, $\vec{E}$ and $\vec{P}$) are vectors. For simplicity, let us assume that $\vec{E}$ and $\vec{P}$ are in the same direction; then we have

$$dU = TdS - P_0 \, dV + Vd\left(\frac{E^2}{8\pi}\right) + VEdP$$

where P_0 is the pressure, and P and E are the scalar polarization density and electric field strength.

Define the Gibbs free energy function for such a dielectric material as:

$$G = U - TS + P_0 \, V$$
$$dG = dU - TdS - SdT + P_0 \, dV + VdP_0$$

It follows that

$$dG = -SdT + VdP_0 + Vd\left(\frac{E^2}{8\pi}\right) + V\,EdP$$

Define a new parameter (P with a hat ˆ):

$$\widehat{P} = P_0 + \frac{E^2}{8\pi} + EP$$

$$d\widehat{P} = dP_0 + d\left(\frac{E^2}{8\pi}\right) + EdP + PdE$$

or

$$Vd\widehat{P} = VdP_0 + Vd\left(\frac{E^2}{8\pi}\right) + V\,EdP + V\,PdE$$

Thus dG equation can be rewritten as

$$dG = -SdT + Vd\widehat{P} - VPdE$$

That is,

$$G = G(T, \widehat{P}, E)$$

From the dG equation, we have the following Maxwell relations:

$$\left[\frac{\partial(V\,P)}{\partial \widehat{P}}\right]_{T,E} = -\left(\frac{\partial V}{\partial E}\right)_{T,\widehat{P}}$$

Note that because

$$\vec{D} = \vec{\varepsilon} \cdot \vec{E} \quad (\text{or} \quad D = \varepsilon E) \quad \text{and} \quad \vec{D} = \vec{E} + 4\pi\vec{P}, \quad \vec{P} = \chi\vec{E}$$

where χ is the electrical susceptibility and ε is the dielectric constant, therefore, when $E = $ constant,

$$P = \text{constant and } D = \text{constant}$$

and hence, from

$$d\widehat{P} = dP_0 + d\left(\frac{E^2}{8\pi}\right) + EdP + PdE,$$

we have

$$d\widehat{P} = dP_0 \quad \text{at} \quad E = \text{constant}$$

Therefore, the above Maxwell relation becomes:

$$\left[\frac{\partial(VP)}{\partial P_0}\right]_{T,E} = -\left(\frac{\partial V}{\partial E}\right)_{T,\widehat{P}}$$

The term on the right and side of the above equation, $\left(\frac{\partial V}{\partial E}\right)_{T,\widehat{P}}$ is called **the electrostriction**, which means that a change of the volume occurs when an electric field is applied. On the left hand side, (VP) is the total electric polarization, and $\left[\frac{\partial(VP)}{\partial P_0}\right]_{T,E}$ implies that the electric property changes when the applied pressure changes.

Another important Maxwell relation can be derived from

$$dG = -SdT + Vd\widehat{P} - V\,PdE.$$

$$\left[\frac{\partial(V\,P)}{\partial T}\right]_{\widehat{P},E} = \left(\frac{\partial S}{\partial E}\right)_{T,\widehat{P}}$$

Using the chain rule, we have

$$\left(\frac{\partial S}{\partial E}\right)_{T,\widehat{P}} = -\left(\frac{\partial S}{\partial T}\right)_{E,\widehat{P}}\left(\frac{\partial T}{\partial E}\right)_{S,\widehat{P}}$$

Let us also define a molar heat capacity at both E and $\widehat{P} = $ constant as

$$C_{E,\widehat{P}} = T\left(\frac{\partial S}{\partial T}\right)_{E,\widehat{P}}$$

We know $P = $ constant when $E = $ constant. From the definition of $\widehat{P}$, if $\widehat{P} = $ constant also, then, $P_0 = $ constant. Therefore,

$$C_{E,\widehat{P}} = C_{E,P_0}$$

$$\left(\frac{\partial S}{\partial E}\right)_{T,\widehat{P}} = -\frac{C_{E,P_0}}{T}\left(\frac{\partial T}{\partial E}\right)_{S,\widehat{P}}$$

Finally, the second Maxwell relation becomes

$$\left[\frac{\partial(VP)}{\partial T}\right]_{E,P_0} = -\frac{C_{E,P_0}}{T}\left(\frac{\partial T}{\partial E}\right)_{S,\widehat{P}}$$

The term on the left hand side of the above equation is called the **Pyro-Electric effect** that means the thermal-electric effect. The right hand side term is called the **Electro-Caloric effect** that means the change of the temperature by applying an electric field.

1.12 Introduction to Thermodynamic Stability

As we have shown, we can use either the entropy maximum principles or the energy minimum principle to find the equilibrium conditions of a thermodynamic system. So far, we have discussed only about the equilibrium conditions. In this section, we will explore the conditions required for a stable equilibrium.

For an isolated system in a stable equilibrium state, the entropy maximum principle requires

$$dS|_U = 0 \quad \text{Equilibrium condition}$$
$$\Delta S|_U < 0 \quad \text{Stable equilibrium condition}$$

Consider that the entropy function at an equilibrium state is given by

$$S = S(X_1, X_2, \ldots X_n) = S(\{X_i\})$$

where the set of variables X_i specifies an equilibrium state. Applying the Taylor expansion to the entropy function around such an equilibrium state, we have

$$S(\{X_i + \Delta X_i\}) = S(\{X_i\}) + \Sigma\frac{\partial S}{\partial X_i}\Delta X_i + \frac{1}{2}\Sigma\Sigma\frac{\partial^2 S}{\partial X_i \partial X_j}\Delta X_i \Delta X_j + \cdots$$

small variation   equilibrium

from the original   state

equilibrium state

Let $\Delta S = S(\{X_i + \Delta X_i\}) - S(\{X_i\})$.
The Taylor expansion can be rewritten as:

$$\Delta S = \Sigma\left(\frac{\partial S}{\partial X_i}\right)\Delta X_i + \frac{1}{2}\Sigma\Sigma\left(\frac{\partial^2 S}{\partial X_i \partial X_j}\right)\Delta X_i \Delta X_j + \cdots$$

In the derivation below, let us assume that the higher order terms in the above Taylor series are negligible, and we keep only the first-order derivatives and the second-order derivatives in the Taylor series.

As we know, at a stable equilibrium state ($\{X_i\}$), there must be:

$$S|_U = S(\{X_i\})|_U = S_{maximum}|_U$$

and

$$\Delta S = S(\{X_i + \Delta X_i\}) - S(\{X_i\}) < 0$$

From the basic calculus, the above condition requires

Equilibrium condition

$$\sum \left(\frac{\partial S}{\partial X_i} \right) \Delta X_i = 0$$

Stability condition

$$\Delta S|_U = \frac{1}{2} \Sigma \Sigma \left(\frac{\partial^2 S}{\partial X_i \partial X_j} \right) \Delta X_i \Delta X_j < 0.$$

For example, a simple system has

$$S = S(U, V)$$

The explicit form of the stability condition

$$\Delta S = \frac{1}{2} \Sigma \Sigma \left(\frac{\partial^2 S}{\partial X_i \partial X_j} \right) \Delta X_i \Delta X_j < 0$$

is given by

$$\Delta S = \frac{1}{2} \left\{ \left(\frac{\partial^2 S}{\partial U^2} \right) (\Delta U)^2 + \left(\frac{\partial^2 S}{\partial V^2} \right) (\Delta V)^2 + 2 \left[\frac{\partial^2 S}{\partial U \partial V} \right] \Delta U \Delta V \right\} < 0$$

Mathematically, according to the theory of quadratic functions, we can show that $\Delta S < 0$ requires:

$$\left(\frac{\partial^2 S}{\partial U^2}\right)_V < 0$$

$$\left(\frac{\partial^2 S}{\partial V^2}\right)_U < 0$$

$$\left(\frac{\partial^2 S}{\partial U^2}\right)\left(\frac{\partial^2 S}{\partial V^2}\right) - \left(\frac{\partial^2 S}{\partial U \partial V}\right)^2 > 0$$

Using the Maxwell relations, we can find out what these stability conditions imply.

For example, $\left(\frac{\partial^2 S}{\partial U^2}\right)_V < 0 \quad \Rightarrow \quad \frac{-N}{T^2 C_V} < 0, \quad \Rightarrow \quad C_V > 0.$

Similarly, consider that the internal energy function at an equilibrium state is given by

$$U = U(X_1, X_2, \ldots X_n) = U(\{X_i\})$$

where the set of variables X_i specifies an equilibrium state. Applying the Taylor expansion to the energy function around such an equilibrium state and neglecting high order terms, we have

$$U(\{X_i + \Delta X_i\}) = U(\{X_i\}) + \Sigma \frac{\partial U}{\partial X_i} \Delta X_i + \frac{1}{2} \Sigma \Sigma \frac{\partial^2 U}{\partial X_i \partial X_j} \Delta X_i \Delta X_j + \ldots$$

small variation   equilibrium

from the original   state

equilibrium state

Let $\Delta U = U(\{X_i + \Delta X_i\}) - U(\{X_i\})$.

The Taylor expansion can be rewritten as:

$$\Delta U = \Sigma \left(\frac{\partial U}{\partial X_i}\right) \Delta X_i + \frac{1}{2} \Sigma \Sigma \left(\frac{\partial^2 U}{\partial X_i \partial X_j}\right) \Delta X_i \Delta X_j + \cdots$$

According to the energy minimum principle, the stable equilibrium conditions in terms of energy are given by

$$dU|_S = 0 \quad \text{Equilibrium condition}$$
$$\Delta U|_S > 0 \quad \text{Stable equilibrium condition}$$

The equilibrium condition requires:

$$\Sigma \left(\frac{\partial U}{\partial X_i}\right) \Delta X_i = 0$$

The stable equilibrium condition requires:

$$\Delta U = \frac{1}{2} \Sigma \Sigma \left(\frac{\partial^2 U}{\partial X_i \, \partial X_j} \right) \Delta X_i \, \Delta X_j > 0$$

The above equation is a quadratic function of real variables, we can further translate it as

$$\Delta U = \Sigma \Sigma \, a_{ij} \, \Delta X_i \, \Delta X_j = \{\Delta X_i\}^T [A] \{\Delta X_j\} = \mathbf{X^T A X} > 0$$

where

$$\mathbf{A} = [a_{ij}] \quad \text{and} \quad a_{ij} = \left(\frac{\partial^2 U}{\partial X_i \, \partial X_j} \right)$$

$\mathbf{X} = \{\Delta X_i\}$ (row vector), $\mathbf{X}^T$ is the transpose row vector.

For a positive quadratic function (ΔU), it requires that

$$a_{ii} = \frac{\partial^2 U}{\partial X_i^2} > 0$$

and the determinant

$$\begin{vmatrix} a_{11} \cdots a_{1\kappa} \\ \cdots \cdots \cdots \\ a_{\kappa 1} \cdots a_{\kappa\kappa} \end{vmatrix} > 0 \quad \text{for all } k = 1, 2, \ldots n$$

For example: a simple system has $U = U(S, V)$. The a_{ii} terms in the stability conditions are (using the Maxwell relations):

$$\left. \frac{\partial^2 U}{\partial S^2} \right|_V = \left. \frac{\partial T}{\partial S} \right|_V = \frac{T}{NC_v} > 0, \quad C_V > 0$$

$$\left. \frac{\partial^2 U}{\partial V^2} \right|_S = -\left. \frac{\partial P}{\partial V} \right|_S = \frac{1}{VK_s} > 0, \quad K_S > 0$$

The determinant term is

$$\left(\frac{\partial^2 U}{\partial S^2} \right) \left(\frac{\partial^2 U}{\partial V^2} \right) - \left(\frac{\partial^2 U}{\partial S \partial V} \right)^2 = \frac{T}{NC_v} \frac{1}{VK_s} - \left[\left(\frac{TV\alpha}{NC_p} \right) \left(\frac{-1}{VK_s} \right) \right]^2 > 0.$$

1.13 Phase Change and Clapeyron Equation

In this section, we will discuss the phase change phenomena involved in multi-component and multi-phase systems, such as a solid–liquid-vapor three-phase system. The phase changes examined here are the so-called first-order phase change, because **these phase changes involve latent heat Δh and the specific volume change Δv**. Why do we call this kind of phase change the first-order phase change? We will explain it in a late section.

Chemical potential is the key parameter in the studies of phase change. Let us consider a multi-component, two-phase (liquid–vapor) system in a piston-cylinder arrangement, as illustrated in the figure below. The system undergoes phase change and reaches an equilibrium at T = constant and P = constant. We use the Gibbs potential to model this system.

$$G = G_V + G_L = \sum \mu_{vi} N_{vi} + \sum \mu_{Li} N_{Li}$$

$$dG = \sum \mu_{vi} dN_{vi} + \sum \mu_{Li} dN_{Li}$$

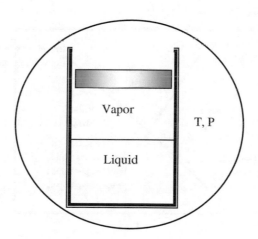

Because the total mass of this system is constant,

$$N_{vi} + N_{Li} = \text{constant}$$

$$dN_{vi} = -dN_{Li}$$

$$dG = \sum (\mu_{Li} - \mu_{vi}) dN_{Li}$$

At equilibrium, the Gibbs potential should be minimum, i.e.,

$$dG = 0.$$

It follows:

$$\mu_{Li} = \mu_{vi} \quad (i = 1, 2, \ldots r)$$

This is the phase change equilibrium condition between a multi-components liquid phase and a multi-components vapor phase.

Because the equilibrium condition of this phase change requires to minimize the total Gibbs potential of the system, that is

$$dG < 0$$

If $(\mu_{Li} - \mu_{vi}) > 0$, it follows $dN_{Li} < 0$.

This means: mass transfers from the phase of higher chemical potential μ (the liquid phase in this case) to the phase of lower chemical potential μ (the vapor phase in this case). That is, the liquid molecules are transferred into the vapor phase, when $\mu_{Li} > \mu_{vi}$.

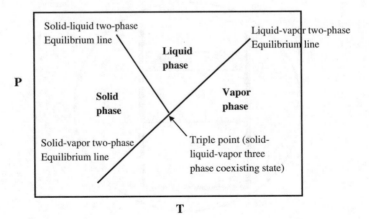

The phase equilibrium states can be presented in terms of a phase diagram. For example, the phase diagram of water is illustrated above. Each line in this diagram represents two-phase equilibrium states or two-phase coexisting states. It should be noted that the curves on the phase diagram are empirically measured. Thermodynamics can predict only the slope of these curves.

Clapeyron Equation

Consider a single-component, liquid–vapor two-phase system at equilibrium.

Because $\mu_L = \mu_v$ and $d\mu_L = d\mu_v$.

and we know

$$d\mu_L = -s_L dT + v_L dP$$

and

$$d\mu_V = -s_V dT + v_V dP$$

We have $(s_V - s_L)dT = (v_V - v_L)dP$

or

$$\frac{dP}{dT} = \frac{s_V - s_L}{v_V - v_L}$$

It should be realized that dP/dT in the above equation is the slope of the Liquid–vapor two-phase equilibrium curve in the phase diagram illustrated above.

Because $\mu = u - Ts + Pv = h - Ts$ and $\mu_L = \mu_v$,

We have $(h - Ts)_L = (h - Ts)_V$

or

$$h_V - h_L = T(s_V - s_L)$$

By definition, the latent heat of evaporation is the difference between the enthalpy of the saturated vapor and the enthalpy of the saturated liquid,

$$L = h_V - h_L$$

Therefore, from the above equation, we have

$$(s_V - s_L) = \frac{L}{T}$$

Finally,

$$\frac{dP}{dT} = \frac{L}{T(v_V - v_L)} \cong \frac{L}{T\,v_V} \qquad \text{(Note here } v_L \ll v_V\text{)}.$$

This equation is called the **Clapeyron Equation**.

The Clapeyron Equation indicates the dependence of equilibrium pressure on temperature for liquid–vapor two-phase equilibrium.

If we assume that the vapor phase is an ideal gas, i.e., $Pv = RT$,

$$\frac{dP}{dT} = \frac{LP}{RT^2} \qquad \rightarrow \qquad \frac{dP}{P} = \frac{L}{RT^2}dT$$

We can show, after integration,

$$P = C \, \exp\left[-\frac{L}{RT}\right] \qquad C = P_0 \exp[-L/RT_0]$$

where T_0 and P_0 are the temperature and pressure of the triple point.

Example: At one atmosphere pressure, i.e., P = 101.41 kPa, water boils at T = 100 °C = 373.15 K. The latent heat of evaporation is $L = 2256.4$ kJ/kg, the specific volume of water vapor is $v_v = 1.6720$ m^3/kg. From the Clapeyron equation, we have.

$$\frac{dP}{dT} = \frac{L}{Tv_V} = \frac{2256.4}{373.15 \times 1.672} = 3.616$$

The above equation may be approximated as:

$$\Delta T \approx \frac{\Delta P}{3.616}$$

If the top of a mountain is 995 m above the sea level, and the air pressure at this mountain top is about 90 kPa, the pressure drop in comparison with the pressure at the sea level (P = 101.41 kPa) is 11.41 kPa. Using the above equation, it follows

$$\Delta T \approx \frac{\Delta P}{3.616} = \frac{-11.41}{3.616} = -3.15\,°C$$

This means that, if you boil water at the top of this mountain, the water will boils at about 96.8 °C, instead of 100 °C.

Following a similar derivation, for the solid–liquid two-phase equilibrium, we can show:

$$\frac{dP}{dT} = \frac{L_{melting}}{T(v_L - v_S)}$$

where $L_{melting}$ is the latent heat of ice melting. For water, $v_S = v_{ice} > v_L$, therefore,

$$\frac{dP}{dT}\bigg|_{ice-liquid} = \frac{L_{melting}}{T(v_L - v_S)} < 0$$

Example: Let us consider water as an example. Water freezes at 0 °C (273.15 K) and 1 atm pressure. At 0 °C (273.15 K) and 1 atm pressure, the latent heat of ice melting is $L_{melting} = 334$ kJ/kg, the specific volume of ice is $v_{ice} = 0.00109$ m3/kg, and the specific volume of liquid water is $v_{liquid} = 0.00100$ m^3/kg.

$$\frac{dP}{dT} = \frac{L_{melting}}{T(v_L - v_S)} = \frac{334}{273.15 \times (0.00100 - 0.00109)} = -13586.3$$

The above equation may be approximated as:

$$\Delta T \approx \frac{\Delta P}{-13586.3}$$

From this equation, we see that, unless you apply enormous pressure (e.g., several hundreds of atmosphere pressure), the ice-water equilibrium temperature or the melting temperature of ice will not change appreciably.

1.14 Chemical Potentials

As we discussed previously, the chemical potential is a key parameter in determining the phase equilibrium and mass transfer. In the following we will first introduce some explicit forms of the chemical potentials for simple and idealized systems. As will be demonstrated later on, these simple explicit forms of the chemical potentials are very useful tools in determining the equilibrium conditions of coexisting phases. We will also analyze the boiling point and the freezing point of ideal (dilute) solutions by using the expressions of chemical potentials in the next section.

Chemical potential of incompressible, pure liquids

Consider a single component, incompressible liquid. From the Gibbs–Duhem equation,

$$d\mu = -sdT + vdP$$

we have

$$\left[\frac{\partial \mu}{\partial P}\right]_T = v$$

For an incompressible liquid, the specific volume v is a constant. Integration of the above equation while keeping T constant gives:

$$\mu_L(T, P) = v_L P + f(T)$$

where f (T) is the integration constant, because the above integration is carried out at a given temperature T, therefore, it is a function of temperature T.

Let P_∞ be a reference pressure at the temperature T. The above equation leads to

$$\mu_L(T, P_\infty) = v_L P_\infty + f(T)$$

From this equation, we can determine the unknown function $f(T)$ as

$$f(T) = \mu_L(T, P_\infty) - v_L P_\infty$$

Finally, the chemical potential of the incompressible, pure liquid at a given state (T, P) can be expressed as:

$$\mu_L(T, P) = \mu_L(T, P_\infty) + v_L(P - P_\infty)$$

In the above, the subscript L indicates the liquid.

Chemical potential of pure ideal gases

Consider a single component ideal gas. From the ideal gas law,

$$Pv = RT$$

where R is the universal gas constant ($R = 8.314$ kJ/kmol K). From the Gibbs–Duhem equation,

$$d\mu = -sdT + vdP$$

we have

$$\left[\frac{\partial \mu}{\partial P}\right]_T = v = \frac{RT}{P}$$

Integrating the above equation while keeping T constant yields

$$\mu_g(T, P) = RT \ln P + g(T)$$

where the subscript g denotes gas, and g (T) is the integration constant.

Let P_α be a reference pressure at the temperature T. The above equation leads to

$$\mu_g(T, P_\infty) = RT \ln P_\infty + g(T)$$

From this equation, we can determine the unknown function $g(T)$ as

$$g(T) = \mu_g(T, P_\infty) - RT \ln P_\infty$$

Finally, the chemical potential of a pure ideal gas at a given state (T, P) can be expressed as:

$$\mu_g(T, P) = \mu_g(T, P_\infty) + RT \ln \frac{P}{P_\infty}$$

Chemical potential of dilute solutions

Consider a dilute solution of two components, and use the subscript 1 to denote the solvent and the subscript 2 to denote the solute. Let the mole fraction be x, defined as

$$x = \frac{N_2}{N_1 + N_2}$$

where N_1 and N_2 are the mole numbers of the solvent and the solute, respectively. For a dilute solution, $x \ll 1$, and $N_2 \ll N_1$, therefore, $x \cong \frac{N_2}{N_1}$.

Consider a dilute solution obeys Raoult's law:

$$P_2 = x P^*$$

Here P_2 is the partial vapor pressure of the solute above the solution, x is its mole fraction of the solute and P^* (T) is the equilibrium vapor pressure of the pure solute at the same temperature.

Realizing that the chemical potential of the solute in the liquid phase is equal to the chemical potential of the solute in the vapor phase at equilibrium, and assuming that the vapor phase behaviors as an ideal gas, we have

$$\mu_2(T, P) = \mu_2(T, P_\infty) + RT \ln \frac{P_2}{P_\infty}$$

Substituting P_2 by using Raoult's law yields:

$$\mu_2(T, P) = \mu_2(T, P_\infty) + RT \ln \frac{xP^*}{P_\infty} = \mu_2(T, P_\infty) + RT \ln \frac{P^*}{P_\infty} + RT \ln x$$

Group the 1st two terms together

$$\psi(T) = \mu_2(T, P_\infty) + RT \ln \frac{P^*(T)}{P_\infty}$$

The chemical potential of the solute in the dilute solution is given by:

$$\mu_2(T, P) = \psi(T) + RT \ln x$$

where $\psi(T)$ is a property of the pure solute at the given state (T).

Consider the vapor of the solvent is an ideal gas.

$$\mu_1(T, P) = \mu_1(T, P_\infty) + RT \ln \frac{P_1}{P_\infty}$$

where P_1 is the partial vapor pressure of the solvent in the vapor phase.

In a mixture of gases, each gas component has a partial pressure which is the assumed pressure of that gas if it alone occupied the volume of the gas mixture at the same temperature. The partial pressure of a gas is the pressure exerted by a gas component in the volume occupied by the gas mixture.

The vapor phase is a mixture, containing both the solvent vapor and the solute vapor. The partial vapor pressure of the solvent is given by:

$$P_1 = \left(\frac{N_1}{N_{total}} \right) P_{total} = \left(\frac{N_{total} - N_2}{N_{total}} \right) P_{total} = (1 - x) P_{total}$$

Let $P_{total} = P_\infty$,

$$\mu_1(T, P) = \mu_1(T, P_\infty) + RT \ln \frac{(1 - x) P_\infty}{P_\infty} = \mu_1(T, P_\infty) + RT \ln(1 - x)$$

Because, when $x \ll 1$.

$$\mathrm{Ln}(1 - x) \approx -x$$

finally, the chemical potential of the solvent can be expressed as:

$$\mu_1(T, P, x) = \mu_1^0(T, P) - xRT$$

where $\mu_1^0(T, P)$ is the chemical potential of **the pure solvent** at the given state *(T, P)*, R is the universal gas constant, T is the given temperature, and x is the mole fraction of the solute in the dilute solution.

Example: Consider a composite system consisting of two subsystems, as illustrated in the figure below. In subsystem A, there is a dilute aqueous solution consisting of water and a salt (NaC1). There is only water in subsystem B. However, the water is present in two phases (i.e., liquid and vapor) in subsystem B. Assume that the total volume and the total mass of the system are constant. The system is surrounded by a thermal reservoir. The constraints for the combined system (the system and the reservoir) are:

$$U + U^R = \text{constant}$$
$$N_i = \text{constant}, \quad i = 1, 2, ...r$$
$$V_{Total} = \text{constant}$$

The internal constraints are

(1)   The subsystems are separated by a rigid partition so that the volumes of the subsystems are constant, i.e., $V_A = \text{constant}$, $V_B = V_{BL} + V_{BV} = \text{constant}$.
(2)   The partition is semi-permeable, and only water molecules can pass through the partition, i.e., $N_{AW} + N_{BW} = \text{constant}$.

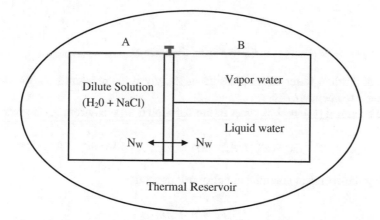

In the above, the subscripts A and B stand for the subsystem A and subsystem B; the subscript W stands for water; and the subscripts L and V stand for liquid and vapor.

In a previous section, we have demonstrated that the equilibrium conditions for this system are the following:

$$P_B^L = P_B^V$$
$$\mu_{AW} = \mu_{BW}^L = \mu_{BW}^V$$

As seen from these equations, it seems that the equilibrium conditions did not reveal any information about the pressure in the subsystem A. Now, we wish to determine the pressure in the subsystem A, P_A.

Solution: First, since the subsystem A and the subsystem B are separated by a rigid partition, there is no volume exchange between the two subsystems, and hence it seems that the pressure of the subsystem A is independent of the subsystem B. However, it should be realized that the only connection between subsystem A and subsystem B is the exchange of water molecules through the semi-permeable partition. Therefore, we will use the chemical equilibrium condition

$$\mu_{AW} = \mu_{BW}^L = \mu_{BW}^V$$

as the starting point.
    Because

$$\mu_{AW} = \mu_{BW}^L$$

note that the chemical potential of pure liquid water in the subsystem B is:

$$\mu_{BW}^L = \mu_W^0(T, P_B^L) = \mu_W^0(T, P_\infty)$$

where

$$P_B^L = P_\infty = P_{saturation@T}$$

is the equilibrium water vapor pressure at the given temperature (i.e., the thermal reservoir's temperature).

The chemical potential of water in the solution (i.e., subsystem A) is given by:

$$\mu_{AW} = \mu(T, P_A, x) = \mu_W^0(T, P_A) - RTx = \mu_W^0(T, P_\infty) + v_W(P_A - P_\infty) - RTx$$

Using the chemical potential equality relationship,

$$\mu_{AW} = \mu_{BW}^L$$

we have

$$\mu_W^0(T, P_\infty) + v_W(P_A - P_\infty) - RTx = \mu_W^0(T, P_\infty)$$

This results in

$$v_W(P_A - P_\infty) = RTx$$

That is,

$$P_A = \frac{RTx}{v_W} + P_\infty$$

where v_W is the specific volume of the liquid at the given temperature T, R is the universal gas constant, x is the mole fraction of the solute in the solution, and

$$P_\infty = P_B^L = P_{saturation@T}$$

The pressure in the subsystem A is determined in this way. Clearly, one can see from the above equation,

(1) P_A is larger than the pressure in the subsystem B, P_∞, and
(2) P_A increases with the mole fraction, x, of the salt in the solution.

One may think about what will happen if the partition separating the two subsystems is allowed to move. How will this affect the pressures in subsystem A and in subsystem B?

If the above system is a biological cell, the separating membrane may have a limited strength to sustain a high pressure difference (at a high x) across the membrane. Let us evaluate the pressure difference across the cell membrane.

Consider a cell suspended in an electrolyte solution, as shown in the figure below. The mole fraction of the electrolyte inside the cell is x_{cell}, and the mole fraction of the electrolyte outside the cell is x_{sol}.

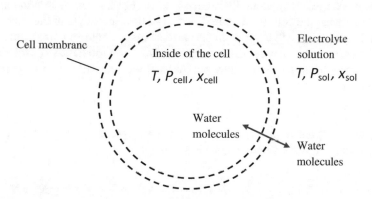

Since the cell membrane allows water molecules to pass through, we will use the chemical equilibrium condition:

$$\mu_{w-cell} = \mu_{w-solution}$$

or

$$\mu_{w-c} = \mu_{w-s}$$

to find the correlation between the pressure inside the cell and the pressure outside the cell.

The chemical potentials of water molecules of the electrolyte solutions inside the cell and outside the cell are given by:

$$\mu_{w-c} = \mu_{w-c}(T, P_c, x_c) = \mu_w^o(T, P_c) - RTx_c = \mu_w^o(T, P_\infty) + v_w(P_c - P_\infty) - RTx_c$$

$$\mu_{w-s} = \mu_{w-s}(T, P_s, x_s) = \mu_w^o(T, P_s) - RTx_s = \mu_w^o(T, P_\infty) + v_w(P_s - P_\infty) - RTx_s$$

In the above equations, the subscript c stands for the cell; the subscript s stands for the solution outside the cell. Inserting these equations into the chemical equilibrium condition yields:

$$v_w(P_c - P_\infty) - RTx_c = v_w(P_s - P_\infty) - RTx_s$$

$$(P_c - P_s) = \frac{RT}{v_w}(x_c - x_s)$$

From the above equation, we see clearly that the pressure difference across the membrane (also called the osmotic pressure), $(P_c - P_s)$, is proportional to difference of the mole fractions of the electrolyte across the membrane, $(x_c - x_s)$. If the mole fraction difference is small, the pressure difference is small. If the mole fraction difference is large, the pressure difference is large. For a given mole fraction of the electrolyte inside the cell, if the mole fraction of the electrolyte in the surrounding liquid is low, $(x_c - x_s)$ will be large. In the extreme case, the surrounding liquid is pure water, i.e., $x_s = 0$, this may result in a pressure difference too large to be tolerated by the membrane. That is why one may see that a blood cell will burst in a pure water environment.

Home work

1. A two-component gas is enclosed in a closed rigid cylinder. The fundamental equation of a two-component gas is given by:

$$S = NA + NR \ln \frac{U^{3/2}V}{N^{5/2}} - N_1 R \ln \frac{N_1}{N} - N_2 R \ln \frac{N_2}{N} \qquad N = N_1 + N_2$$

where A is a constant, R is the gas constant, N_1, N_2 are the mole numbers of component 1 and component 2, respectively; N is the total mole number of the gas; S and U are entropy and internal energy, respectively. The cylinder with a volume of 10 L is divided into two chambers of equal volume by a diathermal rigid membrane. The membrane is permeable to the 1st component only. Chamber A initially has $N_{1A} = 0.5$, $N_{2A} = 0.75$, $T_A = 300$ K. Chamber B initially has $N_{1B} = 1.0$, $N_{2B} = 0.5$, $T_B = 250$ K. After the equilibrium is established, what are the values of T, N_{1A}, N_{1B}, P_A and P_B?

2. Consider a piston-cylinder arrangement that contains a vertical, small diameter tube as shown in the figure below. The system is in contact with a thermal reservoir. The tube contains a dilute solution with a non-volatile solute and the mole fraction is $x < < 1$. The liquid phase and the vapor phase outside the tube consist of only one component, the solvent. The wall of the tube is permeable only to the solvent. Neglecting the surface tension effect, assuming that vapor pressure is constant everywhere in the cylinder, and the density is constant within each phase. Find

(1) Equilibrium conditions.
(2) The height of the liquid column in the tube above the liquid–vapor interface outside the tube when the system is in equilibrium (hint: consider the chemical potential equilibrium condition at the position of the liquid–vapor interface across the tube).

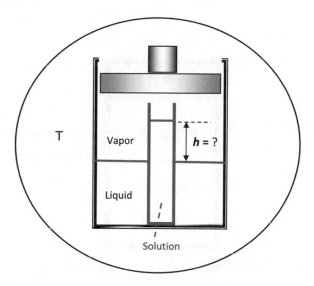

1.15   Boiling Temperature and Freezing Temperature of Dilute Solutions

Question: At one atmosphere pressure, can you make water boil at a temperature higher than 100 °C? Can you make water freeze at a temperature well below 0 °C? How and why?

Question: Two kettles have the same amount of water and are put on a stove to make boiling water. If one puts a bit of salt in one kettle, which kettle will boil first?

Question: Do you know why snow salt can melt snow?

Let us see if we can answer these questions by using the expressions of chemical potentials given in the last section. Consider a dilute solution containing a solute in equilibrium with its vapor phase. The liquid and the vapor are enclosed in a piston-cylinder arrangement and in contact with a pressure reservoir, as illustrated below. Consider the solute is non-volatile so that the vapor phase contains only the solvent molecules. From the Gibbs–Duhem equation, the differential form of the chemical potential of the vapor phase is given by:

$$d\mu_1^v = -s_1^v dT + v_1^v dP = -s_1^v dT \qquad \text{as} \qquad dP = 0$$

The chemical potential of the solvent in the solution (liquid phase) is given by:

$$\mu_1^L(T, P, x) = \mu_1^{0L}(T, P) - RTx$$

and

$$d\mu_1^L = d\mu_1^{0L} - d(RTx) = -s_1^L dT + v_1^L dP - RxdT - RTdx$$

In the above, the subscript 1 and the subscript 2 represent the solvent and the solute, respectively; the superscript L and v stand for the liquid phase and the vapor phase, respectively; R is the universal gas constant, and x is the mole fraction of the solute in the solution.

Because $dP = 0$, the differential form of the chemical potential of the solvent in the solution is reduced to:

$$d\mu_1^L = -s_1^L dT - RxdT - RTdx$$

At the phase equilibrium, we must have:

$$\mu_1^L = \mu_1^v \quad \text{or} \quad d\mu_1^L = d\mu_1^v$$

This leads to:

$$-s_1^L dT - RxdT - RTdx = -s_1^v dT$$

The above equation can be rearranged as

$$(s_1^v - s_1^L)dT = RxdT + RTdx$$

Note that the latent heat of evaporation is

$$L = T(s_1^v - s_1^L),$$

thus,

$$\frac{L}{T}dT = RxdT + RTdx$$

$$\left(\frac{L}{T} - xR\right)dT = RTdx$$

$$\frac{dT}{dx} = \frac{RT}{\frac{L}{T} - xR}$$

It should be realized that we are dealing with the liquid–vapor phase equilibrium, the temperature involved in the above equations is the liquid–vapor two phase equilibrium temperature, or the boiling temperature,

$$T = T_{bp} = T_{vap\text{-}liq\ equilibrium}$$

where the subscript bp stands for boiling point temperature.

Therefore, we may replace T by T_{bp} in the above equation, i.e.,

$$\frac{dT_{bp}}{dx} = \frac{RT_{bp}}{\frac{L}{T_{bp}} - xR}$$

Generally, $\frac{L}{T_{bp}} >> xR$, thus the above equation can be approximated as

$$\frac{dT_{bp}}{dx} = \frac{RT_{bp}^2}{L} > 0$$

This equation shows the dependence of the boiling temperature of a dilute solution on the mole fraction of the solution. Since the right-hand side of the equation is positive, it implies that T_{bp} **increases with the increase of** x. In other words, an increase in the amount of the solute or impurity in a solution will increase the boiling temperature of that solution. For example, at 1 atm pressure, pure water boils at 100 °C; however, tap water (with impurity) will boil at a temperature slightly higher than 100 °C.

For freezing and melting phenomena, we can derive the dependence of freezing temperature T_f on the solute mole fraction in a similar way. We can show

$$\frac{dT_f}{dx} = -\frac{RT_f^2}{L_{hm}} < 0$$

where L_{hm} is the latent heat of melting. It should be noted that there is a negative sign on the right-hand side of this equation. Because the right-hand side of the equation is negative, it implies that the freezing temperature T_f **decreases with the increase of x**. In other words, an increase in the amount of the solute or impurity in a solution will lower the freezing temperature of that solution. For example, at one atmosphere pressure, pure water freezes at about 0–4 °C; however, if some solute such as alcohol (e.g., methanol, or ethylene glycol) is added to water to form a solution, the solution will freeze at a temperature lower than the normal freezing temperature of water.

Take water at 1 atm pressure as an example. $L = 40{,}626$ kJ/kmol, $L_{hm} = 6030.87$ kJ/kmol, the boiling point and the freezing point of pure water are: $T^0{}_{bp} = 373$ K, and $T^0{}_f = 273$ K, respectively.

$$\frac{dT_{bp}}{dx} = \frac{RT_{bp}^2}{L} = \frac{8.314 \times (373)^2}{40626} = 28.47$$

$$\Delta T_{bp} \cong 28.47 \times \Delta x$$

If we add some solute into the water and change the solute mole fraction from initially zero to $\Delta x = 10\% - 0 = 10\%$, we will have $\Delta T_{bp} \cong 2.85\,°C$. That is, the boiling temperature of the solution will be 2.85° higher than that of the pure water.

Similarly,

$$\frac{dT_f}{dx} = -\frac{RT_f^2}{L_{hm}} = -\frac{8.314 \times (273)^2}{6030.87} = -102.8$$

$$\Delta T_f \cong -102.8 \times \Delta x$$

If we add some solute into the water and change the solute mole fraction from initially zero to $\Delta x = 10\% - 0 = 10\%$, we will have $\Delta T_f \cong -10.28\,°C$. That is, the freezing temperature of the solution is approximately 10.3° lower than that of the pure water.

Home Work

(1) A pure liquid is in equilibrium with its vapor phase initially at T_0 and P_0. They are placed in a piston cylinder arrangement and in contact with a thermal reservoir (T_0).

   (a) Prove that a thermodynamic potential function must be minimum in order for equilibrium to exist.

   (b) A small amount of nonvolatile solute is added to the liquid to form a dilute solution (i.e., the mole fraction of the solute $x \ll 1$). Assume that the vapor is an ideal gas. Find the relationship between the original and the new equilibrium vapor pressures as a function of the mole fraction of the solute (hint: you may use the original vapor pressure as a reference pressure in the chemical potential function).

(2)   The boiling point of a liquid is defined as the temperature T_{bp}. Assume a liquid phase is in contact with its vapor and both are contained in a cylinder that is closed by a freely moving piston. The cylinder is surrounded by a pressure reservoir.

(a)   If the liquid contains a non-volatile solute component, derive and show what conditions must the intensive properties satisfy in order for equilibrium to exist?

(b)   Assume the liquid is an ideal aqueous solution with one solute component and the mole fraction is $x \ll 1$. Derive the expression of the boiling temperature.

(c)   If initially the cylinder contains only pure water at its boiling point and then a non-volatile solute is added to the liquid. What will be the direction of mass transport at the liquid–vapor interface? What will be the direction of energy transport at the cylinder-reservoir boundary?

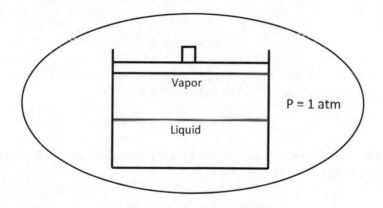

1.16   Gibbs Phase Rule

The equilibrium states of a multi-component, multi-bulk-phase system obey the Gibbs phase rule. For a given number of coexisting phases, the phase rule predicts the number of degrees of freedom or the number of independent intensive variables required to describe the equilibrium state of the system. In other words, for a given number of independent variables, the phase rule predicts the maximum number of coexisting phases.

The Gibbs phase rule is given by

$$f = r + 2 - n \quad (\text{or } n = r + 2 - f)$$

where f is the number of degrees of freedom or the number of independent intensive variables, r is the number of independent chemical components in the system, and n is the number of coexisting phases in the system.

Generally, when a multiphase system is in equilibrium, every phase in the system must reach equilibrium. However, equilibrium of a multiphase system requires additional equilibrium conditions between the phases to be satisfied.

Now let us see how to derive the Gibbs phase rule. Because the number of degrees of freedom is different for different systems, the general way to derive the phase rule (or to count for the degrees of freedom) is:

> The Number of degrees of freedom
>
> = The number of variables − The number of constraints

First, let us consider a simple, single phase:

According to the postulate I, the equilibrium state of such a simple and single phase can be completely characterized by $(r + 2)$ independent extensive variables, for example,

$$(S, V, N_1, \ldots N_r),$$

where r is the number of independent chemical components in the system. Correspondingly, there are $(r + 2)$ intensive variables,

$$(T, P, \mu_1, \ldots \mu_r)$$

However, not all these intensive variables are independent from each other; they are related by the Gibbs-Duhelm equation:

$$SdT - VdP + \sum N_i d\mu_i = 0$$

Therefore, one of these intensive variables can be expressed in terms of others. Only $(r + 1)$ intensive variables are independent for a simple, single phase.

Now, consider a system consisting of r components and n coexisting phases in a mutual equilibrium state. Because each phase has $(r + 1)$ independent intensive variables, and there are n coexisting phases, the total number of intensive variables is

$$n(r + 1).$$

Because the n coexisting phases are in a mutual equilibrium state, the intensive variables are constrained to satisfy the following conditions:

Thermal equilibrium conditions

$$T^\alpha = T^\beta = \cdots = T^n \qquad (n-1) \text{ equations}$$

Mechanical equilibrium conditions

$$P^\alpha = P^\beta = \cdots = P^n \qquad (n-1) \text{ equations}$$

Chemical equilibrium conditions

$$\mu_i^\alpha = \mu_i^\beta = \cdots = \mu_i^n \qquad r(n-1) \text{ equations}$$

$$(i = 1, 2, \ldots r)$$

Therefore, the total number of constraint equations for the $n \, (r+1)$ intensive variables is

$$(n-1)(r+2)$$

Thus, the number of degrees of freedom can be calculated as

$$f = n(r+1) - (n-1)(r+2) = r + 2 - n$$

Example 1: Consider a two-phase (e.g., liquid–vapor) system has one component only. $r = 1, n = 2$. According to the Gibbs phase rule, $f = r + 2 - n = 1$. This implies that there is only one independent intensive variable, and once it is chosen, all other properties are functions of this intensive variable. For example, for liquid water in equilibrium with its vapor, if we choose temperature as the independent intensive variable, the equilibrium pressure is determined by the temperature, $P = P(T)$.

Example 2: Consider a two-phase (e.g., liquid–vapor) system has two components (e.g., a solvent and a solute). $r = 2, n = 2$. Gibbs phase rule predicts $f = r + 2 - n = 2$. This implies that two independent intensive variables are required to describe the equilibrium state. For example, $P = P(T, x)$, where x is the mole fraction, $x = N_2/(N_1 + N_2)$.

Example 3: Consider a three-phase (e.g., ice-liquid–vapor) system has one component. $r = 1, n = 3$. Gibbs phase rule predicts $f = r + 2 - n = 0$. This implies that such a three phase co-existing equilibrium state can exist in only a fixed point in the phase diagram, the triple point O. The three phase co-existing system cannot change.

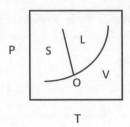

It should be noted that the above phase rule is valid only for simple, bulk phase systems under the following conditions: (1) There is no chemical reaction. (2) The system has no interactions with external fields. (3) There are no surface/interfacial effects.

1.17   Introduction to High-Order Phase Change

Previously, we mentioned first-order phase change. The characteristics of the first-order phase change are the latent heat and the specific volume change, i.e., $\Delta h \neq 0$, and $\Delta v \neq 0$. Liquid–vapor phase change and ice-liquid phase change of water are typical examples of this type of phase changes.

However, there exist phase changes without latent heat and specific volume change. For example, the phase transition between liquid Helium I and liquid Helium II. Helium becomes liquid at 4.2 K and remains liquid form from 4.2 K to absolute zero. However, at $T = 2.2$ K, liquid Helium undergoes a phase transition, i.e., becomes liquid Helium II. The liquid Helium II has very different properties from liquid Helium I. For example, liquid Helium II has extremely high heat conductivity and essentially zero viscosity. Interested readers may find other references about Helium's phase change in libraries.

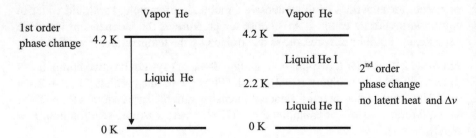

Thermodynamically, different orders of phase change can be summarized as follows:

First order phase change

$$\mu^\alpha = \mu^\beta,$$

That is, the chemical potentials of the two phases (α and β) are equal, at equilibrium.

But: $s^\alpha \neq s^\beta$ and $v^\alpha \neq v^\beta$.

That is, other properties of the two phases are different in value, for instance, the specific entropies and the specific volumes of the two phases are not equal.

From the Gibbs–Duhem equation, we have

$$s = -\left(\frac{\partial \mu}{\partial T}\right) \quad \text{and} \quad v = \left(\frac{\partial \mu}{\partial P}\right)$$

We see that for the first-order phase change, the chemical potentials are equal, but the first-order derivatives of the chemical potentials are not equal. Clearly, that is why we have latent heat and specific volume change:

$$L = T \Delta s \neq 0 \quad \text{and} \quad \Delta v \neq 0$$

That is why we call this type of phase changes as the first-order phase change.

2nd order phase change

$$\mu^\alpha = \mu^\beta, \quad s^\alpha = s^\beta \quad \text{and} \quad v^\alpha = v^\beta$$

For the second order phase change, the chemical potentials and the first-order derivatives of chemical potentials are equal in all phases, and hence there are no latent heat $L = T\Delta s$ and no specific volume change Δv.

However, properties such as C_p, α and k are not equal in different phases.

$$C_P^\alpha \neq C_P^\beta, \quad \alpha^\alpha \neq \alpha^\beta \quad \text{and} \quad k^\alpha \neq k^\beta$$

Recall

$$C_P = T\left(\frac{\partial s}{\partial T}\right) = -T\left(\frac{\partial^2 \mu}{\partial T^2}\right)$$

$$\alpha = \frac{1}{v}\left(\frac{\partial v}{\partial T}\right) = \frac{1}{v}\left(\frac{\partial^2 \mu}{\partial T \partial P}\right)$$

$$k = \frac{1}{v}\left(\frac{\partial v}{\partial P}\right) = \frac{1}{v}\left(\frac{\partial^2 \mu}{\partial P^2}\right)$$

That is, the 2nd-order derivatives of the chemical potentials are not equal. That is why we call this type of phase changes as the 2nd-order phase change.

In general, the nth-order phase change will have:

$$\mu^\alpha = \mu^\beta \qquad \text{and} \qquad \left(\frac{\partial^{n-1}\mu}{\partial x^{n-1}}\right)^\alpha = \left(\frac{\partial^{n-1}\mu}{\partial x^{n-1}}\right)^\beta \quad (n = 1, 2, \ldots)$$

However,

$$\left[\frac{\partial^n \mu}{\partial x^n}\right]^\alpha \neq \left[\frac{\partial^n \mu}{\partial x^n}\right]^\beta$$

Chapter 2
Modelling Homogeneous and Heterogeneous Systems

Abstract So far, we have learned how to model "simple systems" and find equilibrium conditions. A simple system is defined as a system that is homogeneous, isotropic, uncharged, not subject to external interactions (e.g. electric, or gravitational fields), and has no surface or boundary effects. In this chapter, we will show how to establish a thermodynamic model for homogeneous non-simple systems, and heterogeneous non-simple systems, and how to find their equilibrium conditions.

The fundamental equations for a simple system are given by:

$$S = S(U, V, N_1, \ldots \ldots N_r)$$

$$U = U(S, V, N_1, \ldots \ldots N_r)$$

A fundamental equation contains all thermodynamic information of the system. Applying the entropy maximum principle or energy minimum principle to the fundamental equation, we can derive the thermodynamic equilibrium conditions of a given system. Therefore, **the key to model a thermodynamic system is to establish the fundamental equation**.

However, the fundamental equations as shown above are not applicable to many thermodynamic systems that do not satisfy all conditions of a simple system. Therefore, in this chapter, we will show how to establish a thermodynamic model for two types of systems: homogeneous non-simple systems, and heterogeneous non-simple systems.

For a homogeneous system, the properties of the system are uniform throughout the system. The fundamental equation generally can be expressed as:

$$U = U(\text{a set of extensive variables})$$

It should be understood that an extensive variable represents a property of the whole system. Therefore, in order to use extensive variables to characterize the system, the system must be homogeneous, and the values of the properties of the system are uniform throughout the system.

In order to establish the specific fundamental equation for a given homogeneous system, we must understand the physics and chemical processes involved in the system so that we know what extensive variables should be used to characterize the system.

Generally, to characterize a system, we need several types of extensive variables:

(1)   Thermodynamic variable, S (entropy).
(2)   Chemical variables, Ni (mole number), $i = 1, 2, ... r$.
(3)   Mechanical variables, V (volume) for a simple fluid phase or an ideal solid phase; A (surface area) for a simple surface or interface phase; ε_{ij} (strain tensor) when elasticity must be considered.
(4)   Electrical variables, such as N_i (mole number of the charged spices).
(5)   ......

For simple fluid systems and simple solid systems, the set of extensive variables used in the fundamental equations includes S, U, V and N_i as we have shown in Chap. 1. In this chapter, we will demonstrate how to choose different extensive variables to model two example systems: simple elastic solid and simple electrolyte solutions.

With respect to heterogeneous systems, such as a fluid system in a gravitational field or a centrifugal field, because the properties of such a system are not uniform, changing from location to location, we cannot use the extensive variables to characterize the whole system. Instead, we must use.

(1)   the local properties,
(2)   the local equilibrium approximation and
(3)   integration.

to model such a non-uniform system. In this chapter, we will demonstrate how to apply this approach to example systems: systems in gravitational field and in centrifugal field.

2.1   Simple Elastic Solid

An elastic solid is a material that can resist deforming forces and can return to its original size and shape when the forces are removed. When dealing with an elastic solid, we must consider the relative deformation of the material under forces (such as normal stress and shear stress). For example, an elastic band has a length L. If one applies a force to stretch it from the two ends of the elastic band, the length of the elastic band increases and becomes $(L + \Delta L)$. The ratio $(\Delta L \ / \ L)$ is the relative deformation of this elastic band. Generally, such a **relative deformation is called strain**. The strain produced in a body due to tensile force or compressive force or shear force is called the tensile strain, or compressive strain or shear strain, respectively. **The relative change in the volume of a body to its original volume is**

called the volumetric strain. To model a simple elastic solid, the volumetric strain will be used as the mechanical variable, instead of the volume V.

Generally, for a 3-D strain tensor (as the solid is a 3-D object), there are 9 components, as show below:

$$\varepsilon = \begin{bmatrix} \varepsilon_{11} & \varepsilon_{12} & \varepsilon_{13} \\ \varepsilon_{21} & \varepsilon_{22} & \varepsilon_{23} \\ \varepsilon_{31} & \varepsilon_{32} & \varepsilon_{33} \end{bmatrix}$$

For simplicity, we will consider a simple elastic solid. **For such a simple elastic solid, the strain tensor will not only remain the same from position to position in the solid, but also be symmetric**. "Symmetric tensor" means the tensor components are identical across the diagonal line of the matrix, i.e.,

$$\varepsilon_{12} = \varepsilon_{21}, \qquad \varepsilon_{13} = \varepsilon_{31}, \qquad \varepsilon_{23} = \varepsilon_{32}$$

Therefore, among the 9 components, only 6 of them are independent, i.e.,

$$\varepsilon = \begin{bmatrix} \varepsilon_1 & \varepsilon_4 & \varepsilon_5 \\ \varepsilon_4 & \varepsilon_2 & \varepsilon_6 \\ \varepsilon_5 & \varepsilon_6 & \varepsilon_3 \end{bmatrix}$$

Knowing the above strain tensor for this case, the volumetric strain is the initial unconstrained volume multiplying the strain tensor. The fundamental equation of this simple, elastic solid system can be written as:

$$U = U(S, V_0\varepsilon_1, \ldots, V_0\varepsilon_6, N_1, \ldots, N_r)$$

where V_0 is the unconstrained volume (i.e., the volume before deformation), and ε_i ($i = 1, 2, 3, 4, 5, 6$) is the strain tensor component.

The differential form of the fundamental equation is given by

$$dU = TdS + \sum_{i=1}^{6} \sigma_i V_0 d\varepsilon_i + \sum_{j=1}^{r} \mu_i dN_i$$

where

$$\sigma_i = \left(\frac{\partial U}{\partial \varepsilon_i}\right)_{S, \varepsilon_j, N_i}$$

is the stress tensor component, corresponding to the strain tensor component, ε_i.
From the above fundamental equation, the Euler equation is:

$$U = TS + \sum_{i=1}^{6} \sigma_i V_0 \varepsilon_i + \sum_{j=1}^{r} \mu_i N_i$$

Also, from the above fundamental equation, we can derive the following Gibbs–Duhem equation:

$$SdT + \sum_{i=1}^{6} V_0 \varepsilon_i d\sigma_i + \sum_{j=1}^{r} N_i d\mu_i = 0$$

For a single component elastic solid with a constant mass (i.e., $N =$ constant), the Gibbs–Duhem equation is given by:

$$SdT + \sum_{i=1}^{6} V_0 \varepsilon_i d\sigma_i = 0$$

This implies

$$\sigma_i = \sigma_i(T)$$

and

$$\left[\frac{\partial \sigma_i}{\partial T} \right]_{\sigma_{j \neq i}} = -\frac{S}{V_0 \varepsilon_i}$$

or

$$S = -V_0 \varepsilon_i \left[\frac{\partial \sigma_i}{\partial T} \right]_{\sigma_{j \neq i}}$$

This means that the entropy of the simple elastic solid can be determined by measuring the dependence of the stress on temperature.

The Helmholtz free energy of the simple elastic solid is given by:

$$F = U - TS = \sum_{i=1}^{6} \sigma_i V_0 \varepsilon_i + \sum_{j=1}^{r} \mu_i N_i$$

$$dF = -SdT + \sum_{i=1}^{6} \sigma_i V_0 d\varepsilon_i + \sum_{j=1}^{r} \mu_i dN_i$$

An important Maxwell relation can be obtained from the above equation:

$$\left[\frac{\partial \sigma_i}{\partial \varepsilon_j} \right]_{T, N_i, \varepsilon_{k \neq j}} = \left[\frac{\partial \sigma_j}{\partial \varepsilon_i} \right]_{T, N_i, \varepsilon_{k \neq i}}$$

This is a correlation between different strain and stress tensor components. Here $C_{ij} = \left[\frac{\partial \sigma_i}{\partial \varepsilon_j} \right]_{T, N_i, \varepsilon_{k \neq j}}$ is called the isothermal elastic stiffness coefficient.

The Gibbs free energy of the simple elastic solid is given by:

$$G = U - TS - \sum_{i=1}^{6} \sigma_i V_0 \varepsilon_i = \sum_{j=1}^{r} \mu_i N_i$$

$$dG = SdT - \sum_{i=1}^{6} V_0 \varepsilon_i d\sigma_i + \sum_{j=1}^{r} \mu_i dN_i$$

Another important Maxwell relation can be obtained from the above equation:

$$\left[\frac{\partial \varepsilon_i}{\partial \sigma_j} \right]_{T,N_i,\sigma_{k \neq j}} = \left[\frac{\partial \varepsilon_j}{\partial \sigma_i} \right]_{T,N_i,\sigma_{k \neq i}}$$

where $K_{ij} = \left[\frac{\partial \varepsilon_i}{\partial \sigma_j} \right]_{T,N_i,\sigma_{k \neq j}}$ is called the isothermal elastic compliance coefficient.

Consider a system consisting of two simple elastic solids in connected with each other. The total internal energy is the sum of the internal energy of the two subsystems.

$$dU_T = dU_1 + dU_2$$

The differential form of the fundamental equation is

$$dU = TdS + \sum_{i=1}^{6} \sigma_i V_0 d\varepsilon_i + \sum_{j=1}^{r} \mu_i dN_i$$

Consider the system is an isolated system under a constant temperature, the total entropy is constant, $dS = d(S_1 + S_2) = 0$, and the mass is fixed, $dN_i = 0$. Therefore, the differential of the total internal energy of the system ($U = U_1 + U_2$) becomes:

$$dU_T = \sum_{i=1}^{6} \sigma_{i1} V_{0\,1} d\varepsilon_{i1} + \sum_{i=1}^{6} \sigma_{i2} V_{0\,2} d\varepsilon_{i2}$$

In the above, the subscripts 1 and 2 represent the first elastic solid and the second elastic solid, respectively.

Consider a simple case where

$$d\varepsilon_{i1} = 0, \quad \text{and} \quad d\varepsilon_{i2} = 0, \quad i = 2, 3, 4, 5, 6$$

$$\varepsilon_{11} + \varepsilon_{12} = \text{constant}, \quad \text{or} \quad d\varepsilon_{11} = -d\varepsilon_{12}$$

That is, the strain (relative deformation) is not zero only in the 1–1 direction, and the strains are zero in all other directions.

The equilibrium condition (the energy minimum condition) requires:

$$dU_T = -\sigma_{11} V_{0\,1} d\varepsilon_{12} + \sigma_{12} V_{0\,2} d\varepsilon_{12} = [\sigma_{12} V_{0\,2} - \sigma_{11} V_{0\,1}] d\varepsilon_{12} = 0$$

$$\sigma_{12} V_{0\,2} = \sigma_{11} V_{0\,1}$$

This is the mechanical equilibrium condition of the two connecting elastic solids.

2.2 Simple Electrolyte Solution Systems

Many science and engineering processes involve electrolytes. An electrolyte system may be viewed as a solution having a lot of particles with electric charge (e.g., ions). We are interested in the equilibrium conditions of such an electrolyte system under the influence of an external electrical field.

For a simple electrolyte solution system, the fundamental equation in energy form is given by:

$$d\overline{U} = dU + dU_e$$

where $\overline{U}$ is the total internal energy of the electrolyte solution; $U = U(S, V, N_1, ..., N_r)$ is the internal energy of the electrolyte solution without considering the electric field effects; and U_e is the electrical part of the total internal energy, due to the interaction of the charged particles with the applied electrical field. It can be shown that

$$dU_e = \psi\, dQ$$

where ψ is the electrical potential of the electrical field, and Q is the total charge of ions.

$$Q = F \sum Z_i N_i$$

where F is the Faraday constant; Z_i is the electro-valence of the ith ionic species; and N_i is the mole number of the ith ionic species. Thus,

$$dU_e = F\psi \sum Z_i dN_i$$

Recall

$$dU = TdS - PdV + \sum \mu_i dN_i$$

Finally, the differential form of the fundamental equation can be written as:

$$d\overline{U} = TdS - PdV + \sum (\mu_i + Z_i F\psi) dN_i$$

or

$$d\overline{U} = TdS - PdV + \sum \overline{\mu}_i dN_i$$

where

$$\overline{\mu}_i = \left(\frac{\partial \overline{U}}{\partial N_i}\right)_{S,V,N_j} = \mu_i + Z_i F \psi$$

.

is called the **electro-chemical potential** of the ionic component i. From the above differential equations, we see that the fundamental equation of the electrolyte system is given by:

$$\overline{U} = U(S, V, \{N_i\})$$

The Euler equation is given by:

$$\overline{U} = TS - PV + \sum \overline{\mu}_i N_i$$

The Gibbs free energy or potential function is:

$$G = \overline{U} - TS + PV = \sum \overline{\mu}_i N_i$$

$$dG = -S\,dT + V\,dP + \sum \overline{\mu}_i dN_i$$

For a multi-component ($i = 1, 2, \ldots\ldots r$) and multi-phase ($k = \alpha, \beta, \ldots\ldots \lambda$) electrolyte system, the general equilibrium conditions can be obtained by using, for example, the energy minimum principle, i.e.,

$$d\overline{U}_{Total} = \sum d\overline{U}_k = 0$$

and by using the constraints:

$$\sum dS_k = 0, \quad \sum dV_k = 0, \quad \text{and} \quad \sum dN_{i,k} = 0$$

It can be shown that the equilibrium conditions are:

$$T^\alpha = T^\beta = \ldots = T^\lambda$$

$$P^\alpha = P^\beta = \ldots = P^\lambda$$

$$\overline{\mu}_i^\alpha = \overline{\mu}_i^\beta = \ldots = \overline{\mu}_i^\lambda \quad (i = 1, 2, \ldots\ldots r)$$

In the following, we will show that the electrochemical potential equilibrium condition can be used to determine ion distribution of an electrolyte solution in electric field. Let us consider a charged solid surface surrounded by an electrolyte solution in an equilibrium state. It should be realized that the charged solid surface generates an electric field in the surrounding solution. Because the solid surface is charged, it will attract the counter-ions (ion with opposite charges) in the electrolyte solution to a region close to the charged surface, and repel the co-ions (ions with the same charges) from the this region. The charge on the solid surface and the balancing charge (counter-ions) in a thin liquid layer close to the surface is the so-called **electric double layer**, as illustrated in the figure on the next page.

At equilibrium, the electrochemical potential of the ions must be constant everywhere, i.e.,

$$\text{grad } \widetilde{\mu}_i = 0$$

where electrochemical potential is $\overline{\mu}_i = \mu_i + Z_i F \psi$. Note that the Faraday constant is $F = e\, N_A$, where e is the elementary charge of an electron, N_A is the Avogadro number (6.022×10^{23}). If we divide the electrochemical potential by the Avogadro number, it becomes the electrochemical potential per ion,

$$\overline{\mu}_i = \mu_i + Z_i e \psi$$

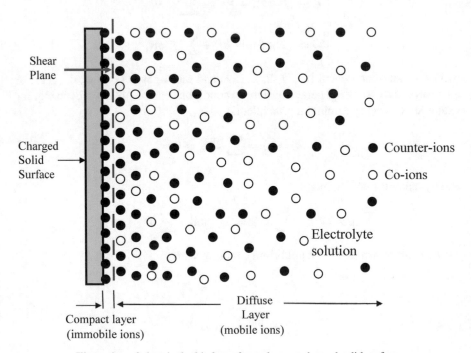

Illustration of electric double layer formed near a charged solid surface

Illustration of electric double layer formed near a charged solid surface. The above equilibrium condition can be further written as:

$$\text{grad } \mu_i = -z_i e \text{ grad } \psi$$

The above equation is the equilibrium condition and shows the balance of forces on ions between the electrical force (as indicated by the gradient of electrical potential ψ on the right-hand side of the above equation) and the diffusion (mass transfer) force (as indicated by the gradient of chemical potential on the left-hand side of the above equation.

For simplicity, let us consider a flat solid surface inserting vertically into an infinitely larger electrolyte solution. For such a one-dimensional system, the above equation can be re-written as:

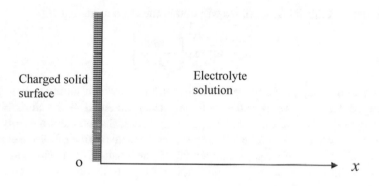

$$\frac{d\mu_i}{dx} = -z_i e \frac{d\psi}{dx}$$

As the chemical potential of an ion species is given by:

$$\mu_i = \mu_i^\infty + k_b T \ln n_i$$

where n_i is the ion number density per unit volume of type i ions (i.e., positive ions $n+$ or negative ions n^-), and μ_i^∞ is a function of T and P only. We have

$$\frac{d \ln n_i}{dx} = \frac{1}{n_i} \frac{dn_i}{dx} = -\frac{z_i e}{k_b T} \frac{d\psi}{dx}$$

Let us integrate this equation from a point in the bulk solution (far away from the charged solid surface) to a point in the region very close to the charged solid surface (i.e., the electric double layer region), and using the following boundary conditions:
In the bulk solution:

$$x = \infty, \quad \psi = 0, \quad \text{and} \quad n_i = n_i^\infty, \text{or} \quad n_+^\infty = n_-^\infty = n_\infty,$$

where n_i^∞ is the bulk ionic number density per unit volume of type i ion in a position far away from the charged solid surface (i.e., not influenced by the electric field of the charged solid surface); n_+^∞ is the bulk ionic number density per unit volume of positively charged ions, and n_-^∞ is the bulk ionic number density per unit volume of negatively charged ions. This means that at positions infinitely far away from the charged solid surface the bulk solution is electrically neutral or has zero net charge (i.e., $n_+^\infty = n_-^\infty$).

In the region very close to the charged solid surface (i.e., inside the electric double layer):

$$x = x, \quad \psi = \psi, \quad \text{and} \quad n_i = n_i.$$

Carrying out this integration, we will obtain the Boltzmann equation:

$$n_i = n_i^\infty \exp\left[-\frac{z_i e \psi}{k_b T}\right]$$

Boltzmann equation describes the distribution of the ion number density near the charged solid surface. Recall ψ is the electrical potential of the electrical field generated by the charged solid surface, $z_i e \, \psi$ represents the interaction energy of the electric field with ions, $k_b T$ reflects the energy of thermal motion of ions. Boltzmann equation indicates that, for a given temperature T, the ion number density is a function of the potential ψ of the electrical field that in turn is a function of the distance from the charged solid surface.

According to the theory of electrostatics, the relationship between the electrical potential ψ and the local net charge density per unit volume ρ_e at any point in the solution is described by the Poisson equation:

$$\nabla^2 \psi = -\frac{\rho_e}{\varepsilon \varepsilon_0}$$

where ε is the relative dielectric constant of the solution, and ε_0 is the dielectric permittivity of vacuum.

Using the Boltzmann distribution equation, the number concentration of the type-i ions in a symmetric electrolyte (symmetric electrolyte means ionic valence ratio z_i: $z_j = 1$, for example, for NaCl, z_i: $z_j = 1{:}1$.) solution is of the form:

$$n_i = n_{i\infty} \exp\left(-\frac{z_i e \psi}{k_b T}\right)$$

where $n_{i\infty}$ and z_i are the bulk ionic concentration and the valence of type-i ions, respectively, e is the charge of an electron, κ_b is the Boltzmann constant, and T is the absolute temperature.

For example, for positive ions,

$$n_+ = n_{+\infty} \exp\left[\frac{z_+ e\psi}{k_b T}\right]$$

and for negative ions,

$$n_- = n_{-\infty} \exp\left[\frac{z_- e\psi}{k_b T}\right]$$

In the bulk electrolyte, far away from the charged solid surface, the electrolyte solution is electrically neutral. That is,

$$n_{+\infty} = n_{-\infty} = n_\infty$$

The net volume charge density ρ_e is proportional to the concentration difference between symmetric cations and anions, via

$$\rho_e = ze(n_+ - n_-) = -2zen_\infty \sinh\left(\frac{ze\psi}{k_b T}\right)$$

where n_+ and n_- are the number density of the positive ions and negative ions, respectively.

Substituting this expression of net charge density into the Poisson equation leads to the well-known Poisson–Boltzmann (P–B) equation.

$$\nabla^2 \psi = \frac{2zen_\infty}{\varepsilon\varepsilon_0} \sinh\left[\frac{ze\psi}{k_b T}\right]$$

By defining the Debye–Huckel parameter $k^2 = \frac{2z^2 e^2 n_\infty}{\varepsilon_o \varepsilon\, k_b T}$ and the non-dimensional electrical potential $\Psi = \frac{ze\psi}{k_b T}$, the Poisson–Boltzmann equation can be re-written as:

$$\nabla^2 \Psi = k^2 \sinh \Psi$$

Generally, by solving this equation with appropriate boundary conditions, the electrical potential distribution $\psi(x)$ of the electric double layer can be obtained. Then using $\psi(x)$, the local charge density distribution $\rho_e(x)$ can be determined.

It should be noted that the Debye–Huckel parameter $k^2 = 2z^2 e^2 n_\infty / \varepsilon\varepsilon_o k_b T$ is independent of the solid surface properties and is determined by the liquid properties (such as the electrolyte's valence and the bulk ionic concentration) only. $1/k$ is normally referred to as the characteristic thickness of electric double layer and is a function of the electrolyte concentration. For example, value of $1/k$ ranges from 9.6 nm at an ionic concentration of 10^{-3} M to 304.0 nm at an ionic concentration of 10^{-6} M for a NaCl or KCl solution. When the ionic concentration is 10^{-6} M, the solution is considered as the pure water. The thickness of the diffuse layer of the electric double layer usually is about three to five times of $1/k$, and hence may be larger than one micron for pure water and pure organic liquids.

Let's see an example of how to calculate $1/k$. Consider pure water at T = 298 K and use the following parameters: $\varepsilon = 78.5$, $\varepsilon_o = 8.85 \times 10^{-12}$ C^2/Nm^2, $e = 1.602 \times 10^{-19}$ C, $k_b = 1.381 \times 10^{-23}$ J/K, and $N_a = 6.022 \times 10^{23}$ /mol. Note that n_∞ is the bulk ionic number concentration and is expressed in terms of the molarity M (mole/liter) by:

$$n_\infty = \left(M\, \tfrac{mol}{L}\right)\left(1000 \tfrac{L}{m^3}\right)\left(N_a \tfrac{1}{mol}\right) = 1000 N_a M$$

Put all the above parameter values into

$$\frac{1}{k} = \left[\frac{\varepsilon \varepsilon_o k_b T}{2z^2 e^2 n_\infty}\right]^{1/2}$$

and we have

$$\frac{1}{k} = \frac{3.04}{z\sqrt{M}} \times 10^{-10} \quad (m)$$

here M is the molarity of a symmetrical $(z{:}z)$ electrolyte.

For example, let $z = 1$, we can calculate the dependence of $1/k$ on molarity M. As seen from the table below, when the bulk ionic concentration increases, more counterions are attracted to the region close to the charged solid surface to neutralize the surface charge. Consequently, the electric double layer thickness is reduced, and it seems like that the electric double layer is "compressed".

M	$1/k$ (nm)
10^{-6}	304.0
10^{-4}	30.4
10^{-2}	3.0

For a flat surface inserted vertically in an infinitely large aqueous solution, if we set the origin of the coordinate system on the wall, as illustrated in the figure below, the Poisson–Boltzmann equation in dimensional form is given by

$$\frac{\partial^2 \psi}{\partial x^2} = \frac{2zen_\infty}{\varepsilon \varepsilon_0} \sinh\left(\frac{ze\psi}{k_b T}\right)$$

with the boundary conditions:

$x = 0$, $\psi = \psi_0 \approx \varsigma$ (i.e., at the solid surface or at the shear plane)

$x = \infty$, $\psi = 0$ (i.e., infinitely far away from the solid surface)

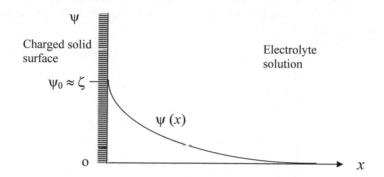

If we assume $\left(\frac{z\,e\psi}{k_b T}\right) < 1$, for example, when $\psi < 25$ mV at 20 °C; and use $sinh(x)$ $\approx x$, The above Poisson–Boltzmann equation is reduced to

$$\frac{\partial^2 \psi}{\partial x^2} = \frac{2zen_\infty}{\varepsilon\varepsilon_0}\left(\frac{z\,e\psi}{k_b T}\right)$$

This simplified equation can be solved under the above–listed boundary conditions. The result is:

$$\psi = \psi_o e^{-kx} = \varsigma\, e^{\ kx}$$

In the above equation, ψ_o is the potential at the solid surface and usually is approximated by a measurable electrokinetic potential at the shear plane, ς, which is called zeta potential, that is, $\psi_o \approx \varsigma$.

Knowing the electrical field distribution, $\psi\,(x)$, we can calculate the distributions of the cations and the anions, and the net charge density distribution near the charged solid surface by using the Boltzmann distribution equation.

$$n_i(x) = n_{i\infty}\exp\left(-\frac{z_i\,e\,\psi\,(x)}{k_b T}\right)\quad (i = +\text{ and }-)$$

$$\rho_e(x) = ze(n_+(x) - n_-(x)) = -2zen_\infty\,\sinh\left(\frac{z\,e\,\psi\,(x)}{k_b T}\right)$$

Homework

Calculate and plot the distributions of the cations and the anions, and the net charge density distribution near a charged solid surface for a KCl solution. The bulk KCl concentration is: 1×10^{-6} M, 1×10^{-4} M, and 1×10^{-2} M. $\varsigma = 25$ mV.

2.3   Systems in Gravitational Field and in Centrifugal Field

Let us consider a moving system subject to a gravitational field and a centrifugal field. For example, a cylinder filled with a solution is spinning around a vertical shaft, as illustrated in the figure below. Such a centrifugal device is often used in industrial separation processes.

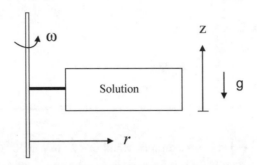

Because the presence of external fields (the gravitational field and the centrifugal field in this case), the properties of the solution are no longer uniform. For example, the pressure and the density of the solution are functions of position, i.e.,

$$P = P(r, z) \quad \text{and} \quad \rho_i = \rho_i(r, z)$$

Obviously, such a system is not a simple system; therefore, we **cannot use** the fundamental equation for a simple system such as

$$U = U(S, V, N_1, \ldots .N_r)$$

to model such a system.

How to Model a System in a Gravitational Field and in a Centrifugal Field

First, we will introduce a so-called "**local equilibrium**" approximation. That is, we **assume, although different elements of a system have different properties, each element itself is a simple system at an equilibrium state, i.e., each element has uniform properties throughout the element.** Therefore, all thermodynamic theories and equations for simple systems are applicable to each element.

For example, the left figure below illustrates a stationary cylinder containing a solution or gas in a gravity field. It is not difficult to understand that different layers in the cylinder along the height direction have different densities and different pressures.

However, a layer with a differential thickness dz at a location z has essentially the same density and the same pressure throughout the layer. Thus we can consider each layer as a simple thermodynamic system.

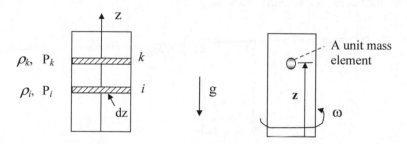

Now let us consider a rotating cylinder containing a fluid in a gravity field. For convenience, let us take a unit mass as an element of the system, as illustrated in the right-hand side figure above. The total energy of the unit mass element is given by:

$$e = u + ke + pe \quad \text{(kJ/kg)}$$

where e is the total energy density per unit mass (kJ/kg); u is the internal energy density (the internal energy for the unit mass element) (kJ/kg); $ke = V^2/2$ (here V is the velocity) is the kinetic energy of this unit mass element (kJ/kg); and $pe = gz$ (here g is the gravity acceleration constant, z is the variable to measure the elevation from ground) is the potential energy of the unit mass element (kJ/kg).

It should be noted that the total energy density e is a function of the position of the unit mass element, i.e., $e = e(r, z)$, because the fluid is subject to both the gravity field and the centrifugal field.

The total energy of the entire system can be obtained by integrating the energy density over the volume of the system.

$$E = \int_V \rho e \, dV = \int_V e' dV$$

where ρ is the local density of the fluid (kg/m³), e is the local total energy density per unit mass (kJ/kg), and $e' = \rho e$ is the **local total energy density per unit volume (kJ/m³)**.

Similarly, the total entropy of the entire system is

$$S = \int_V \rho s \, dV = \int_V s' dV$$

where s is the **local entropy density per unit mass (kJ/kgK)** of the fluid, ρ is the local density of the fluid (kg/m^3), and $s' = \rho s$ is the **local entropy density per unit volume (kJ/m^{3}K)**.

Now, let us consider an isolated fluid system consisting of r components in a gravitational field. The system is stationary.

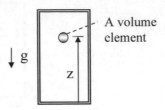

For this system, the total entropy is

$$S = \int_V s' dV$$

and the total energy is given by:

$$E = \int_V e' dV$$

where

$$e' = \rho(u + pe) = \left(\sum n_i M_i\right)(u + gz) = u' + \sum n_i M_i gz$$

$$\rho = \sum n_i M_i$$

$$pe = gz$$

n_i and M_i are the mole density per unit volume (kmol/m^3), and the molar mass of the *ith* component (kg/kmol), respectively.

How to Determine the Equilibrium Conditions

As we have already known that finding thermodynamic equilibrium conditions requires finding the maximum of the entropy function or the minimum of the energy functions. Such a process involves a function of several variables which may be related to each other by one or more constraint equations. For simple systems, these mathematical operations are straight forward as we have demonstrated in the previous chapter. However, for non-uniform systems, the total entropy and the total energy of such a system often involves integration of local properties, and the mathematical analysis to find the maximum of the entropy function or the minimum of the energy functions is relatively complex. Therefore, we need to introduce a **method of undetermined Lagrange multipliers.**

For example, let us consider minimizing a function $z = x^2 + y^2$ where the variables x and y are subject to a constraint equation: $x y - 1 = 0$. An intuitive approach may be to solve the constraint equation first to get:

$$y = 1/x$$

Then substituting y with this equation into $z = x^2 + y^2$ yields:

$$z = x^2 + 1/x^2.$$

Differentiating this equation and setting it to zero (i.e., condition required for minimum) gives:

$$\frac{dz}{dx} = 2x - \frac{2}{x^3} = 0$$

Solving this equation, one can find the minimum positions are $(+1, +1)$ and $(-1, -1)$.

However, the approach used in the above example may not be practical for some more complicated equations. A more general approach is the **method of undetermined Lagrange multipliers,** as outlined below.

Consider minimization or maximization of a function

$$F(x, y \ldots)$$

under a set of constraint equations:

$$C_i(x, y, \ldots) = 0 \qquad (i = 1, \ldots n)$$

First, let us define a new Lagrange function

$$L = F + \sum_{i=1}^{n} \lambda_i C_i$$

where the constants $\lambda_1, \lambda_2, \ldots \lambda_n$ are called the undetermined Lagrange multipliers.

The conditions or positions of the minimization or maximization can be found by solving the following sets of equations:

$$\frac{\partial L}{\partial x} = \frac{\partial L}{\partial y} = \ldots = 0$$

$$C_i(x, y, \ldots) = 0 \qquad (i = 1, \ldots n)$$

Let us use the above example to illustrate this method.
The function is:

$$z = x^2 + y^2$$

The constraint is:

$$x\, y - 1 = 0$$

The first step of using the method of undetermined Lagrange multipliers is to define:

$$L = F + \sum_{i=1}^{n} \lambda_i\, C_i = (x^2 + y^2) + \lambda\,(xy - 1)$$

The set of the equations to be solved are:

$$\frac{\partial L}{\partial x} = 0 \qquad\qquad 2x + \lambda y = 0$$

$$\frac{\partial L}{\partial y} = 0 \qquad\qquad 2y + \lambda x = 0$$

$$C(x, y) = 0 \qquad\qquad x y - 1 = 0$$

Solving this system of equations gives: $\lambda = -2$, and the minimum positions are $(+1, +1)$ and $(-1, -1)$.

Now let us see how to apply the method of undetermined Lagrange multipliers to find the equilibrium conditions for a system in a gravitational field and in a centrifugal field.

First, let us consider an isolated, stationary fluid system consisting of r components in a gravitational field. If this isolated system is in a thermodynamic equilibrium state, we know, from the entropy maximum principle,

$$S = \int_V s'dV \Rightarrow Maximum$$

or

$$\delta S = 0$$

where δ is the mathematical symbol of small variation (students may see textbooks of Calculus of Variations regarding δ), and can be approximated as the first order derivative.

The entropy maximum is achieved under the following constraints:

$$E = \int_V e' dV = \text{constant}$$

or

$$\delta E = 0$$

(i.e., the total energy of an isolated system is constant), and

$$N_i = \int_V n_i dV = \text{constant} \quad i = 1, 2, \ldots, r$$

or

$$\delta N_i = 0$$

(i.e., the total mass of an isolated system is constant).

Mathematically, the task of finding the equilibrium conditions is to maximize the total entropy under these constraint conditions, i.e.,

$$\delta S = 0 \text{ with constraints } \delta E = 0 \text{ and } \delta N_i = 0.$$

Using the undetermined Lagrange Multipliers method as introduced above, this entropy maximization problem can now be formulated as follows:

$$\delta L = \delta (S - \lambda_0 E - \sum \lambda_i N_i) = 0$$

where the Lagrange function L is defined as

$$L = S - \lambda_0 E - \sum \lambda_i N_i$$

where λ_0 and λ_i are the to-be-determined Lagrange multipliers and are constant. The above equation can be further written as:

$$\delta\left[\int (s' - \lambda_0 e' - \sum \lambda_i n_i)dV\right] = 0$$

Recall

$$e' = u' + \sum n_i M_i gz,$$

and move the δ sign inside the integration, the above equation becomes:

$$\int [\delta s' - \lambda_0(\delta u' + \sum M_i g z \,\delta n_i) - \sum \lambda_i \delta n_i]dV = 0$$

The entropy density per unit volume is given by the following function:

$$s' = s'(u', \{n_i\})$$

Similar to

$$ds = \frac{du}{T} - \sum \left(\frac{\mu_i}{T}\right)dn_i$$

we have

$$\delta s' = \frac{\delta u'}{T} - \sum \left(\frac{\mu_i}{T}\right)\delta n_i$$

Substituting $\delta s'$ in the integration by the above equation, we have

$$\int \left[\left(\frac{1}{T} - \lambda_0\right)\delta u' + \sum_{i=1}^{r}\left(-\frac{\mu_i}{T} - \lambda_0 M_i g z - \lambda_i\right)\delta n_i\right]dV = 0$$

Because the integral must be equal to zero for all variations, it follows:

$$\frac{1}{T} = \lambda_0 = \text{constant} \tag{2.1}$$

$$\frac{\mu_i}{T} + \lambda_0 M_i \, g \, z = -\lambda_i = \text{constant} \tag{2.2}$$

These are the equilibrium conditions for the system in our consideration. Combining Eq. (2.1) with Eq. (2.2) yields

$$\lambda_0 \mu_i + \lambda_0 M_i g z = -\lambda_i$$

or

$$\mu_i + M_i g\, z = (-\lambda_i/\lambda_0) = \text{constant} \qquad (2.3)$$

Therefore, the equilibrium conditions become:

$$T = \text{constant} \qquad (2.4)$$

$$\mu_i + M_i gz = \text{constant} \qquad (2.5)$$

From Eq. (2.5), we can easily see that

$$\mu_i - \text{constant} - M_i g\, z$$

or

$$d\mu_i = -M_i g\, dz$$

That is, the chemical potential at equilibrium is no longer a constant; instead, it is a function of the height (the gravitation effect),

$$\mu_i = \mu_i(z)$$

Consider the Gibbs–Duhem equation for a unit volume element,

$$s'dT - dP + \sum n_i d\mu_i = 0$$

Since $T = \text{constant}$,

$$dP = \sum n_i d\mu_i$$

Use
$$d\mu_i = -M_i g dz.$$
We have:

$$dP = \sum n_i(-M_i g\, dz) = -\left(\sum n_i M_i\right) g\, dz$$

Realize that the density of the fluid is

$$\rho = \sum n_i M_i,$$

the above equation becomes:

$$dP = -\rho \, g \, dz$$

Generally, the density of the fluid is a function of the position along the z axis, $\rho = \rho(z)$. Let us consider the following special cases:

(1)   Assume $\rho = \rho_{average} = \overline{\rho} = $ constant, for example, an incompressible liquid. Integrating $dP = -\rho \, g \, dz$ gives:

$$P(z) = P_{z=0} - \overline{\rho} \, g \, z$$

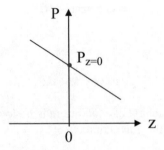

(2)   Assume the fluid is an ideal gas. $\rho = \frac{P}{RT}$

and $dP = -\rho \, g \, dz$ becomes.

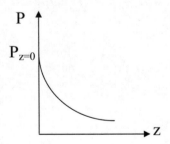

$$\frac{dP}{P} = \frac{-g}{RT} dz$$

Integration yields:

$$P(z) = P_{z=0} \exp\left[\frac{-g \, z}{R \, T}\right]$$

The atmospheric pressure essentially follows the above equation.

Now, let us consider a cylinder filled with a fluid is rotating around a shaft at a speed ω. Generally, such a system can be treated similarly to the systems in the gravity field. Realize that, for a unit mass, the centrifugal field potential, pe_ω, is:

$$pe_\omega(r) = -\tfrac{1}{2}\omega^2 r^2$$

(Compare it with the gravitational potential, $pe_g(z) = gz$),

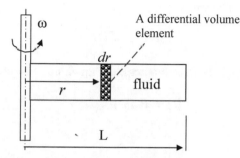

A volume element with thickness dr at a radial distance r from the axis of rotation is shown in the figure above. The total energy of this volume element is:

$$e' = u' + \rho \times pe_\omega(r) = u' + \left(\sum n_i M_i\right)\left(\tfrac{-1}{2}\omega^2 r^2\right)$$

Performing the similar analysis as we did for the system in gravity field, we can show that the equilibrium conditions for a fluid system in a centrifugal field are given by:

$$T = \lambda = \text{constant}$$

$$\mu_i - \tfrac{1}{2}M_i\omega^2 r^2 = \text{constant}$$

or

$$\mu_i = \text{constant} + \frac{1}{2}M_i\omega^2 r^2$$

That is, the chemical potential at equilibrium is no longer a constant; instead, it is a function of the radial distance r (the centrifugal effect), $\mu_i = \mu_i(r)$.

It follows that

$$d\mu_i = M_i\omega^2 r\,dr$$

and from the Gibbs–Duhem equation for a volume element with $dT = 0$

$$\sum n_i d\mu_i = dP$$

we have

$$dP = \sum n_i M_i \, \omega^2 r \, dr$$

Note that $\rho = \sum n_i M_i$.
The above equation becomes:

$$dP = \rho \omega^2 r \, dr$$

If we assume the fluid is an incompressible liquid, i.e., $\rho = $ constant,

$$P(r) = P_{r=0} + \tfrac{1}{2}\rho \omega^2 r^2$$

Home Work

(1)   A fluid is placed in a rigid container and spun as indicated in the figure below. The angular frequency of the rotation is ω (rad/s). The fluid fills the container and has no energy exchange with the surroundings. Gravity effect is negligible in this case.

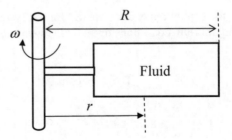

(a)   If the fluid contains more than one component, what conditions must the intensive thermodynamic properties satisfy in order for equilibrium to exist?

(b)   The fluid is a dilute liquid solution of two components (One is dissolved in the other). Assume that the properties of the solute are independent of pressure. The total mole concentration $n = n_1 + n_2 = $ constant (independent of r). Find the mole concentration of the solute as a function of the radial distance, i.e., $n_2 = n_2(r)$.

(2) A cylinder contains a pure, incompressible liquid, as illustrated in the figure below. The cylinder is spun around its longitudinal (the vertical) axis at an angular frequency ω. Neglect any effect due to the presence of vapor, curved liquid–vapor surface and any energy exchange between the liquid and the surroundings. **Do not neglect gravity effect**. Derive the equilibrium conditions and the expression for the pressure as a function of position (r, z) in the liquid phase. Let the pressure at the bottom and the center of the cylinder be P_0.

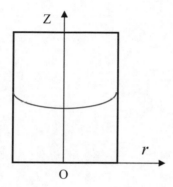

Chapter 3
Thermodynamics of Interfaces and Three-Phase Contact Lines

Abstract Up to now, all the thermodynamics theories we have discussed are the theories for simple three dimensional (3D) bulk phase systems. The size of such a system is measured by volume, and the mechanical work mode for these systems is PdV type of work. One of the conditions for simple systems is that effects of the boundaries of the bulk phases on the equilibrium states are not considered. However, in a broad spectrum of applied science and engineering applications, the bulk phase boundaries, such as an interface between a liquid droplet and its vapor phase, play important roles in determination of the equilibrium states of the system. In this chapter, we will first show how to establish thermodynamic models for 2D surfaces/interfaces and for 1D three-phase contact lines. Then, we will demonstrate how to find the equilibrium conditions of these systems by using the analytical methods explained in Chapter 1 and Chapter 2. The effects of the bulk phase boundaries, i.e., surfaces/interfaces and three-phase contact lines, on the equilibrium conditions will be discussed.

3.1 Introduction to Interfaces and Three-Phase Contact Lines

Up to now, all the thermodynamics theories we have discussed are the theories for simple three dimensional (3D) bulk phase systems. The size of such a system is measured by volume, and the mechanical work mode for these systems is PdV type of work. One of the conditions for simple systems is that effects of the boundaries of the bulk phases on the equilibrium states are not considered. However, in a broad spectrum of applied science and engineering applications, the bulk phase boundaries, such as an interface between a liquid droplet and its vapor phase, play important roles in determination of the equilibrium states of the system. Examples where surface or interface effects are important include (bubble, droplet or ice) nucleation processes, bubble flotation processes used in mineral and oil processing, two-phase (liquid–gas or water–oil) transport phenomena in a porous medium, etc.

The boundaries between immiscible bulk phases include interfaces and three-phase contact lines. **An interface or a surface is the boundary between two immiscible bulk phases. A three-phase contact line is the mutual boundary of three immiscible bulk phases or three surfaces.**

© The Author(s), under exclusive license to Springer Nature Switzerland AG 2022
D. Li, *Analytical Thermodynamics*,
https://doi.org/10.1007/978-3-030-90517-0_3

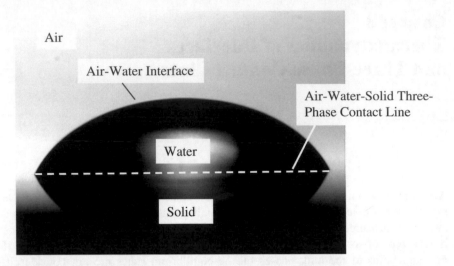

Image of a water drop resting on a flat solid surface

When do we have to consider the effects of these boundaries? Generally, when the dimension of the system is small enough and the energy associated with the surfaces and lines is comparable to the energy associated with the volume of the bulk phase, the effects of the surfaces and three-phase contact lines must be considered.

It should be realized that the systems with important surface or line effects are not the "simple" systems as we defined before. In this chapter, we will show how to apply the general thermodynamic theory we learned before to the two-dimensional (2D) surfaces and the one-dimensional (1D) lines. The objectives in this chapter are to learn.

(1)   How to apply the thermodynamic theory to model systems involving surfaces and lines, and
(2)   How these boundaries will affect the equilibrium conditions.

Interfaces or Surfaces

- What is a surface or an interface?

A water drop hanging from a needle tip

Let us look at the above image of a water drop hanging from the tip of a needle in the air. The profile of the drop seems to indicate that the liquid–air interface is a sharp surface. However, this impression is rather misleading. Generally speaking, when two immiscible, uniform bulk phases are in contact, there is a very thin boundary region between them. The phase properties change rapidly across this thin boundary region, as illustrated in the figure below.

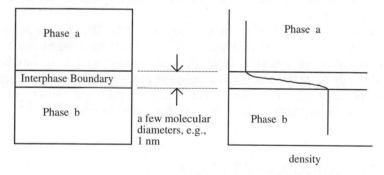

Consider an isolated system consisting of two immiscible bulk phases, *a* and *b*. One of these phases is a liquid and the other phase may be either another liquid or a gas. As illustrated in the figure above, there is a thin interfacial region or inter-phase boundary between these two bulk phases. **Usually the thickness of such an interfacial region is about a few molecular diameters**. The order of magnitude of the thickness of this interfacial region generally is about 10 Å for ordinary liquid-fluid pairs like water and air. Crossing this thin boundary region, the density and other physical properties of the fluids undergo a sharp transition from one bulk phase to the other bulk phase, as illustrated in the figure above. Therefore, **physically, the interface or surface is a three-dimensional (i.e., with a finite thickness), non-homogeneous (i.e., with varying properties) boundary region**.

It is easy to understand that a molecule in a bulk liquid phase is subject to uniform intermolecular forces from all directions. In other words, the forces exerting on a molecule in the bulk liquid phase are balanced from all directions. However, the molecules at the interfacial region experience unbalanced intermolecular forces. This is because, generally, the densities of the bulk phases forming the interface are different, and the molecules on the two sides of the interface may also be different. Thus the attractive intermolecular force acting on a molecule in the interfacial region is stronger from one side of the interface than that from the other side. For example, consider a water–air interface. The density of liquid water is much higher than the density of air. There are more molecules from the liquid water side to interact with a molecule in the interfacial region. There are a fewer molecules from the air side to interact with a molecule in the interfacial region. The net intermolecular force acting on the interfacial molecules is from the liquid water side and tends to pull the molecules into the bulk liquid phase. Therefore, the liquid molecules have a tendency to stay inside the bulk liquid phase, not in the interfacial region. That is why **a liquid-fluid system always tries to minimize its interface area**. If there are no other external influences, a liquid drop always chooses a spherical shape (the spherical shape has the minimum surface area for a given volume). In addition, because the interfacial region has a lower molecular density, the separation distance between molecules is larger than the molecular equilibrium separation distance in the bulk liquid phase. Hence the molecules in the interfacial region experience a stronger attraction force among them. Mechanically, this attraction force manifests itself as a measurable tensile force in all liquid-fluid interfaces; it is called the **surface or interfacial tension**.

As discussed above, **a surface or interface is a thin, 3D, non-homogeneous, boundary layer between two bulk phases. There is a sharp change in the system properties across this boundary layer**. However, as the thickness of the interface is very thin (say, about one nanometer), negligible in comparison with the interface's lateral dimensions (in millimeters or centimeters), we may approximately replace the 3D interfacial region by a 2D mathematical surface (no thickness).

3.2   Thermodynamics of Surfaces

Following the thermodynamics approach we have learned so far, a surface or an interface will be treated as:

1.   A 2D phase as an approximation. This is based on the fact that the thickness of the surface or interface is negligible in comparison with its lateral dimensions.

2. A uniform phase, and all surface properties (U, S, N_1, ... N_r) are considered as the average values over the interfacial thickness. Because the properties of a real interface is non-uniform only in the thickness direction.

In this way, a 3D, heterogeneous interfacial phase is approximated as a 2D, homogeneous, simple thermodynamic phase. All the thermodynamic principles and the equations for simple systems are applicable to such a surface phase. For a 3D bulk phase system, volume V is used to characterize the size of the system, and the volume change is used to characterize the mechanical work (compression or expansion). However, for a 2D surface phase, there is no volume, instead, the surface area A is used to characterize the size of the system, and the surface area change is used to characterize the mechanical work. Therefore, the fundamental equations of a surface phase are given by:

$$S_A = S_A(U_A, A, N_{1A}, \dots N_{rA})$$

and

$$U_A = U_A(S_A, A, N_{1A}, \dots N_{rA})$$

In the above equations, the variable A is the surface area (m^2), and the subscript A indicates the surface phase.

Generally, surfaces or interfaces may not be flat and they are curved. It should be mentioned that there are no curvature variables in these fundamental equations; therefore, these fundamental equations are valid only for moderately curved interfaces where the curvature effects of the interface are negligible. **What is a moderately curved interface?** Generally, when the radius of curvature of an interface is comparable to the thickness of the interface, such an interface is highly curved. If the radius of curvature is at least 100 times greater than the thickness of the interface, such an interface is moderately curved. For a moderately curved interface, the only geometrical variable required to describe the interface is the surface area. However, for a highly curved interface, certain curvature variables are also required to describe the interface, in addition to the surface area. This is because the energy associated with the shape change (such as shape change due to bending or twisting) of the interface is significant part of the total energy of the interface.

From

$$U_A = U_A(S_A, A, \{N_{iA}\}),$$

the differential form of the fundamental equation in terms of energy is given by:

$$dU_A = TdS_A + \gamma \, dA + \sum \mu_i dN_{iA}$$

where two of the intensive properties, the temperature T and the chemical potential μ_i, are defined similarly to that for bulk phases. The only new parameter here is the surface tension γ, defined as:

$$\gamma = \left(\frac{\partial U_A}{\partial A}\right)_{S_A, N_{iA}} \left[mJ/m^2\right] \text{ or } [mN/m]$$

In the last section, we have briefly described how surface tension is generated in the interface. Surface tension as a mechanical force is a physical property of the interface. It has a unit of force per unit length, mN/m (milli-Newton per meter). From the above thermodynamics definition, the surface tension is defined as the change of surface internal energy with respect to the surface area change. Therefore, it is also energy of the surface per unit surface area, with a unit of energy per unit surface area, mJ/m^2 (milli-Joule per square meter).

From the differential form of the fundamental equation, we see that the mechanical work term is given by:

$$dW_A = \gamma \, dA$$

This is called the surface work. The surface tension is therefore considered as the work or energy required creating a unit new surface area.

The order of magnitude of the surface tension ranges from 10 to 72 mJ/m^2 for most liquid-fluid interfaces at room temperature. The term "interfacial tension" is applicable to all liquid–gas, liquid–liquid, solid–gas and solid–liquid interfaces. However, in the literature, people prefer to use the term "surface tension" for liquid–gas (vapor) and solid–gas (vapor) interfaces, and use the term "interfacial tension" for liquid–liquid and solid–liquid interfaces. Generally, solid–fluid interfacial tensions cannot be measured directly. However, there are many methods for measuring the surface tensions of liquid-fluid interfaces. Some typical values of interfacial or surface tensions are listed in Table 3.1.

Recall that the differential form of the fundamental equation in terms of energy for a bulk phase is given by:

$$dU_B = TdS - PdV + \sum \mu_i dNi$$

Table 3.1 Typical values of surface tensions

Liquid	Temperature ($^\circ$C)	Surface tension (mN/m)
Mercury	20	484
Water	20	72.8
Glycerol	20	63.1
Ethylene glycol	20	47.7
n-Octane	20	21.8
Argon	−183	11.9

and the differential form of the fundamental equation in terms of energy for a surface
phase is given by:

$$dU_A = TdS_A + \gamma \, dA + \sum \mu_i dN_{iA}$$

From the above two equations, let us compare the definition of pressure P and the
definition of surface tension γ:

$$-P = \left(\tfrac{\partial U_B}{\partial V}\right)_{S,N_i} \quad [\text{J/m}^3 = \text{N/m}^2]$$

$$\gamma = \left(\tfrac{\partial U_A}{\partial A}\right)_{S_A,N_{iA}} \quad [\text{J/m}^2 = \text{N/m}]$$

We see that the surface tension γ is the 2D analogy to the hydrostatic pressure P
for a 3D bulk phase. That is, γ is a mechanical property of a surface phase while P is
the mechanical property of a volume phase. In fact, if we simply replace $(-P)$ by γ
and V by A, all the previously derived thermodynamic equations for bulk phases will
be transformed into equations for surface phases. The following summarizes some
of the key equations.

The Euler equation for a surface phase is:

$$U_A = TS_A + \gamma A + \sum \mu_i N_{iA}$$

The Gibbs–Duhem equation for a surface phase is:

$$S_A dT + A d\gamma + \sum N_{iA} d\mu_i = 0$$

Helmholtz potential or free energy for a surface phase is:

$$F_A = U_A - TS_A = \gamma A + \sum \mu_i N_{iA}$$

$$dF_A = -S_A dT + \gamma \, dA + \sum \mu_i dN_{iA}$$

Gibbs potential or free energy for a surface phase is:

$$G_A = U_A - TS_A - \gamma A = \sum \mu_i N_{iA}$$

$$dG_A = -S_A dT + A d\gamma + \sum \mu_i dN_{iA}$$

Grand canonical potential or free energy for a surface phase is:

$$\Omega_A = U_A - TS_A - \sum \mu_i N_{iA} = \gamma A$$

The specific grand canonical free energy per unit surface area is:

$$\omega_A = \frac{\Omega_A}{A} = \gamma$$

That is why surface tension γ is also called the specific surface free energy (mJ/m^2).

$$d\Omega_A = -S_A dT + \gamma dA - \sum N_{iA} d\mu_i$$

When $dT = 0$ and $d\mu_i = 0$, we have

$$d\Omega_A = \gamma\, dA \quad \text{or} \quad \left(\frac{\partial \Omega_A}{\partial A}\right)_{T,\mu_i} = \gamma$$

It should be noted that the Gibbs–Duhem equation for a surface

$$S_A dT + A d\gamma + \sum N_{iA} d\mu_i = 0$$

can lead to an important equation, called the Gibbs adsorption equation. Let's consider a liquid-fluid interface in a constant temperature process, i.e., $dT = 0$. The above Gibbs–Duhem equation is reduced to

$$-A d\gamma = \sum N_{iA} d\mu_i$$

or

$$-d\gamma = \sum \frac{N_{iA}}{A} d\mu_i = \sum \Gamma_i d\mu_i$$

where $\Gamma_i = \frac{N_{iA}}{A}$ is called the surface mole density of the *ith* component. For a two-component solution system, $r = 2$, the above equation becomes:

$$-d\gamma = \Gamma_1 d\mu_1 + \Gamma_2 d\mu_2$$

where the component 1 is the solvent and the component 2 is the solute.

From the above equation, we get:

$$\Gamma_2 = -\left(\frac{\partial \gamma}{\partial \mu_2}\right)_{\mu_1}$$

Γ_2 is called the relative surface adsorption of the component 2. It is the surface mole density of the component 2.

At equilibrium, the chemical potential of a given component in the surface phase is the same as the chemical potential of the same component in the bulk phase. Recall that the chemical potential of the solute (component 2) in a dilute solution is given by:

$$\mu_2(T, P) = \psi(T) + RT \ln x$$

where $\psi(T)$ is a material constant at the given temperature; R is the universal gas constant; and x is the mole fraction of the component 2 (solute) in the solution. Using this relation, we have:

$$\Gamma_2 = -\frac{x}{RT}\frac{d\gamma}{dx}$$

This is the Gibbs adsorption equation.

Using this equation, one can determine the relative adsorption Γ_2 on the surface by measuring the surface tension change with respect to the mole fraction of the solute. A typical $\gamma \sim x$ curve is illustrated in the figure below. According to the above Gibbs adsorption equation, the slope of the $\gamma \sim x$ curve is the indication of the relative adsorption of the component 2 at the interface.

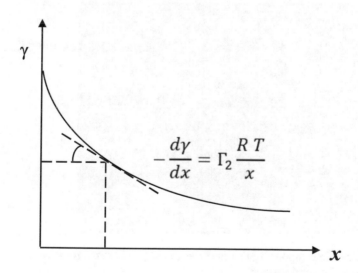

3.3 Thermodynamics of Three-Phase Contact Lines

The mutual boundary of three immiscible bulk phases or the intersection of three interfaces is called the three-phase contact line. For example, as illustrated in the figure below, a three-phase contact line is formed by a sessile drop on a solid substrate surrounded by a vapor phase. As we have already known that each interface is a non-homogeneous interfacial region with a finite thickness, the three-phase contact line that is formed by the intersection of three interfaces is also a non-homogeneous zone with a finite thickness. Material properties changes sharply through this line zone from one bulk phase/interface phase to another. However, the thickness of the three-phase contact zone is negligible in comparison with its length. Therefore, in the surface thermodynamics, similar to the treatment of interfaces, the three-phase contact line is treated as an one-dimensional, uniform linear phase.

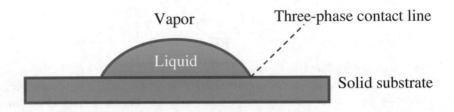

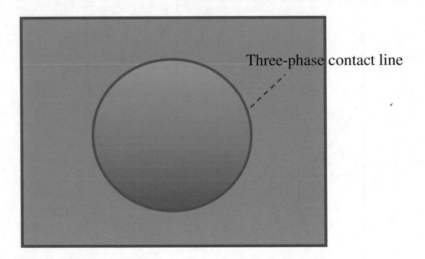

The side view (upper figure) and the top view (lower figure) of a sessile drop resting on a flat solid surface.

The fundamental equation in the energy form for a line phase is given by:

$$U_L = U_L(S_L, L, N_{1L}, \ldots N_{rL})$$

where the variable L is the length of the line phase, and the subscript L indicates the line phase. This equation is valid only for moderately-curved line phases, as no

curvature variables are introduced in the fundamental equation. The differential form
of the fundamental equation is:

$$dU_L = TdS_L + \sigma dL + \sum \mu_i dN_{iL}$$

where the new parameter σ is the line tension, defined by

$$\sigma = \left(\frac{\partial U_L}{\partial L}\right)_{S_L, N_{iL}} \quad [\mu J/m]$$

Comparing the definition of the line tension with the definitions of surface tension
and pressure, we see that σ is the one-dimensional analogy to γ in a 2D surface phase
and P in a 3D bulk phase. Similar to surface tension γ, line tension σ is the mechanical
force operating in the three-phase contact line, it is also the energy required to create
a unit length of the three-phase contact line. There is no consensus on the order of
magnitude of line tension at the present time. For many solid–liquid-vapor systems,
σ has been reported to be of the order of 1 $\mu J/m$ at room temperature.

If we simply replace $(-P)$ by σ and V by L, all the previously derived thermodynamic equations for bulk phases will be transformed into equations for line phases.
For example, the grand canonical free energy of a line phase is given by:

$$\Omega_L = U_L - TS_L - \sum \mu_i N_{iL} = \sigma L$$

and

$$\omega_L = \frac{\Omega_L}{L} = \sigma \quad [\mu J/m]$$

That is why σ is also called the specific free energy of a line phase.

The line tension effects are especially important when the dimension of the
system is very small, for example, in the studies of heterogeneous nucleation. More
discussions will be given in later sections.

3.4 Equilibrium Conditions of Droplets and Bubbles

Consider an isolated system consisting of a liquid droplet surrounded by the liquid's
vapor phase, as illustrated in the figure below.

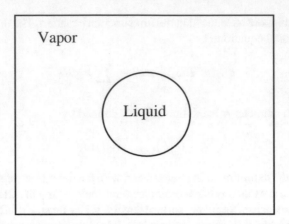

This system consists of two bulk phases: the liquid and the vapor, and one interface phase: the liquid–vapor interface. The total entropy of this composite system is:

$$S = S_L + S_V + S_{LV}$$

where the subscripts, L, V and LV represent the liquid phase, the vapor phase and the liquid–vapor interface phase, respectively. The constraints for this isolated system are the following:

Total energy is constant:

$$U_L + U_V + U_{LV} = \text{constant}$$

Total mass is constant:

$$N_{iL} + N_{iV} + N_{iLV} = \text{constant} \quad (i = 1, 2, \ldots \ldots r)$$

Total volume is constant:

$$V_L + V_V = \text{constant}$$

As we know, at the equilibrium state, the total entropy of this isolated system will be maximum. Applying the entropy maximum principle, $dS = 0$, we have:

$$
\begin{aligned}
dS &= dS_L + dS_V + dS_{LV} \\
&= \tfrac{dU_L}{T_L} + \tfrac{P_L}{T_L} dV_L - \sum \tfrac{\mu_{iL}}{T_L} dN_{iL} \\
&\quad + \tfrac{dU_V}{T_V} + \tfrac{P_V}{T_V} dV_V - \sum \tfrac{\mu_{iV}}{T_V} dN_{iV} \\
&\quad + \tfrac{dU_{LV}}{T_{LV}} - \tfrac{\gamma}{T_{LV}} dA - \sum \tfrac{\mu_{iLV}}{T_{LV}} dN_{iLV}
\end{aligned}
$$

Using the constraints, we have:

$$dU_{LV} = -dU_L - dU_V$$

$$dN_{iLV} = -dN_{iL} - dN_{iV}$$

$$dV_V = -dV_L$$

Furthermore, it is not difficult to understand that the surface area of the droplet depends on the volume of the droplet. Therefore,

$$A = A(V_L)$$

and

$$dA = \left(\frac{dA}{dV_L}\right)dV_L$$

Using these differential constraint relations, the above dS equation becomes:

$$
\begin{aligned}
dS &= dS_L + dS_V + dS_{LV} \\
&= \left(\tfrac{1}{T_L} - \tfrac{1}{T_{LV}}\right)dU_L + \left(\tfrac{1}{T_V} - \tfrac{1}{T_{LV}}\right)dU_V \\
&\quad + \left[\tfrac{P_L}{T_L} - \tfrac{P_V}{T_V} - \tfrac{\gamma}{T_{LV}}\left(\tfrac{dA}{dV_L}\right)\right]dV_L \\
&\quad - \sum\left(\tfrac{\mu_{iL}}{T_L} - \tfrac{\mu_{iLV}}{T_{LV}}\right)dN_{iL} - \sum\left(\tfrac{\mu_{iV}}{T_V} - \tfrac{\mu_{iLV}}{T_{LV}}\right)dN_{iV} \\
&= 0
\end{aligned}
$$

Thus, the equilibrium conditions are:

$$T_L = T_V = T_{LV} \tag{3.4.1}$$

$$\mu_{iL} = \mu_{iV} = \mu_{iLV} \qquad i = 1, 2, \ldots r \tag{3.4.2}$$

$$P_L - P_V = \gamma\left(\tfrac{dA}{dV_L}\right) \tag{3.4.3}$$

As we see from these equations, the thermal and chemical equilibrium conditions, Eqs. (3.4.1) and (3.4.2), are the same as in the cases of bulk phase systems. The mechanical equilibrium condition, however, is changed. Equation (3.4.3) is called the **Laplace Equation of Capillarity**.

It can be shown that for a spherical droplet,

$$\left(\tfrac{dA}{dV_L}\right) = \tfrac{2}{R},$$

where R is the radius of the droplet. Therefore, for a spherical drop, the Laplace equation can be written as:

$$P_L - P_V = \frac{2\gamma}{R}$$

Clearly, because $\frac{2\gamma}{R} > 0$, the liquid pressure (inside the curved liquid–vapor interface) is higher than the vapor pressure (outside the curved liquid–vapor interface), i.e., $P_L > P_V$. This is different from the mechanical equilibrium condition for bulk phases (i.e., $P_L = P_V$), because of the effects of the surface tension γ and the curvature ($1/R$). If either $\gamma = 0$ (without considering the interface) or $R = \infty$ (a flat surface), the Laplace equation will revert to $P_L = P_V$.

If the system is a bubble surrounded by a liquid phase, we can derive the equilibrium conditions by following the above procedure. The mechanical equilibrium condition, i.e., the Laplace equation now is given by:

$$P_V - P_L = \frac{2\gamma}{R}$$

It should be noted that, comparing with the Laplace equation for a liquid drop, the positions of P_L and P_V are exchanged, and R is the radius of the bubble. This means that the pressure inside the bubble is higher than the liquid pressure outside the bubble.

Generally, we can show that

$$\left(\frac{dA}{dV_L} \right) = \frac{1}{R_1} + \frac{1}{R_2} = J$$

where R_1 and R_2 are the principle radii of curvature at a point of a curved surface. As shown in the figure below, two perpendicular planes go through the same point on a curved surface. One plane has a larger radius of curvature, called the principle radius of curvature, R_1. The other plane has a smaller radius of curvature, called the principle radius of curvature, R_2.

$$J = \frac{1}{R_1} + \frac{1}{R_2}$$

is called the mean curvature.

Therefore, the Laplace equation in a more general form is given by:

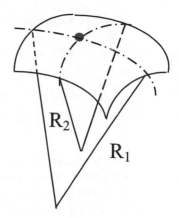

$$\Delta P = \gamma \left(\frac{1}{R_1} + \frac{1}{R_2} \right) = \gamma \cdot J$$

Obviously, for a spherical surface, $R_1 = R_2 = R$, and $J = \frac{2}{R}$.

It should be noted that if the vapor phase is replaced by another fluid (a gas or another immiscible liquid), the Laplace equations derived above remains valid. We only need to replace P_V by P_f (i.e., the fluid pressure) in the Laplace equations.

Home Work

1. As illustrated in the figure below, two bubbles are formed at the ends of a capillary tube in two separate chambers. One bubble is smaller, and the other bubble is larger. Initially, the valve at the middle of the tube is closed, and the two bubbles of different sizes are stable. Assume that the temperature and liquid pressure in the two chambers are the same and are constant. What will happen if the valve is opened? Please use the Laplace equation of capillarity to support your conclusion.

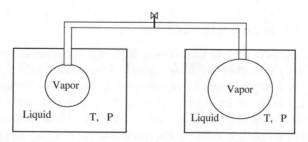

2. A paper-board boat of 50 g is floating on the surface of a still water pond. A slot connects a small hole in the center of the boat to water surface at the back of the boat, as shown below. If one deposits a few drops of oil in the hole, the oil will migrate to the water surface at the back of the boat and the boat will start move forward. Neglect the thickness of the boat.

    (a) What causes the boat to move?
    (b) Assume the boat moves slow enough and the oil spreads over the water surface behind the boat, the surface tension of water is 72 mJ/m^2 and the

surface tension of oil is 36 mJ/m^2. Find the acceleration and velocity of the boat.

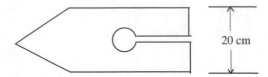

3. A small amount of liquid is deposited into a capillary tube that is placed horizontally, and a liquid column is formed in the capillary tube, as shown in the figure below. The cross-section of the capillary tube is circular and the radius of the capillary tube varies linearly in the length direction. The radii of the two surfaces of the liquid column are r_1 and r_2, respectively. The capillary tube has a finite length, and its two ends are open to the atmosphere.

    (1) Will the liquid column stay in the position? Or move? Show why.
    (2) If it moves, to which direction? What is the final equilibrium configuration? Show why.

    (Hint: Using the Laplace equation).

3.5   Equilibrium Conditions of Sessile Drops

Introduction to Wetting Phenomena

From daily experience, we know that a drop of water deposited onto the surface of a horizontal plastic plate will form a sessile drop. Also, we often see that water can climb up to a certain height in a capillary tube and form a meniscus at the top. In these phenomena, the angle formed between the tangential line of the liquid–vapor interface and the tangential line of the liquid–solid interface at the three-phase contact line is conventionally defined as the **contact angle θ**. Contact angle represents the ability of a liquid to wet a solid surface. It is also sometime referred to as the wetting angle.

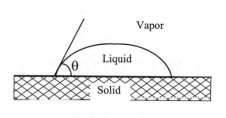

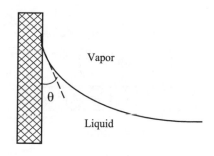

The contact angle defined above is the macroscopic contact angle that can be observed by our eyes or by a regular optical microscope. It should be noted that, at molecular level, the three immiscible bulk phases do not meet in a line; they meet in a zone of small but finite dimensions in which the three interfacial regions merge. Therefore, the microscopic contact angles may be different from the contact angles measured by ordinary instruments, like a goniometer. However, discussion of such microscopic contact angles is beyond the scope of this chapter. For our purpose, the macroscopic contact angles will be considered; and the type of systems studied here consists of three bulk phases: solid, liquid and vapor, and three interface phases: the solid–liquid interface, the solid–vapor interface and the liquid–vapor interface.

The interest in wetting and contact angles is twofold: They play a major role in a number of technological, environmental and biological processes. They are also a manifestation of the surface tension of the solid on which the contact angle is formed.

When the contact angle is larger than 90°, we say that the liquid does not wet the solid surface well, or the solid surface is partially not wettable by the liquid. In the extreme case, when the contact angle is 180°, we say that the solid surface is completely not wettable by the liquid. In this case, the liquid drop will form a sphere with a point contact on the solid surface. On the other hand, when the contact angle is less than 90°, we say that the liquid partially wets the solid surface. In the extreme case, when the contact angle is 0°, we say that the solid surface is completely wetted by the liquid. In this case, the liquid actually forms a film on the solid surface.

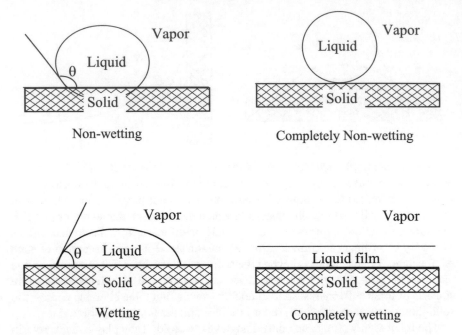

Depending on the applications, the wetting phenomena can also be divided into equilibrium wetting and dynamic wetting. In equilibrium wetting, there is no motion of the solid–liquid-fluid three-phase contact line relative to the solid surface. The contact angle is determined by the intrinsic properties of the solid–liquid-fluid system, such as the surface tensions and the roughness and heterogeneity of the solid surface. In dynamic wetting, the three-phase contact line moves and the contact angle changes generally with time in the moving process. Because of the liquid motion, the liquid viscosity plays an important role in the dynamic wetting processes. If there is no external force involved, such as in the case of spontaneous spreading of a liquid drop on a solid surface, the dynamic wetting is driven by the capillary forces (surface tension) and moves towards the equilibrium wetting configuration. The final contact angle will be the equilibrium contact angle of the system. If the dynamic wetting involves external forces such as pressure and gravity force to drive the bulk liquid flow, the wetting process is more complicated and the contact angle depends on both the system's intrinsic properties and the external forces.

In this chapter, we are interested only in equilibrium wetting and contact angle phenomena. The objectives include (1) to understand the factors determining the contact angles, and (2) to predict the equilibrium contact angle.

Surface Tension and Surface Stress

The wetting phenomena involve solid surfaces. We must understand the basic difference between the solid surfaces and the liquid-fluid interfaces. For most liquid-fluid interfaces, the molecules at the interface have high mobility. That is, despite the shape change of the interface, the molecules can always re-orient themselves to take the minimum energy configuration so that the surface tension remains constant. However, at solid–liquid interfaces, the solid molecules are immobile. When the shape of the solid surface changes, the intermolecular forces change and hence the surface tension and the surface energy are changed. We must know the difference between the surface stress and surface tension.

Generally, surface tension is the work required creating unit area of new surface. Surface stress, however, is associated with the work involved in deforming a surface. Consider a surface of area A is deformed and the corresponding surface area change is dA. The area change dA can be expressed in terms of the change of strain tensor, $\Delta \varepsilon_{ij}$, by

$$dA = A \sum \Delta \varepsilon_{ij} \, \delta_{ij} \quad (i, j = 1, 2)$$

where

$$\delta_{ij} = 1, \quad \text{when } i = j \quad \delta_{ij} = 0, \quad \text{when } i \neq j$$

The amount of work associated with this deformation is given by

$$dW = A \sum g_{ij} \Delta \varepsilon_{ij} \quad (i, j = 1, 2)$$

where g_{ij} is the surface stress tensor.

Recall that the total surface (grand canonical) free energy is γA. Alternatively, the work required for deforming the surface should be equal to the total surface free energy change and can also be expressed as

$$dW = d(\gamma A) = \gamma dA + Ad\gamma = \gamma A \sum \Delta \varepsilon_{ij} \delta_{ij} + A \sum \left(\frac{\partial \gamma}{\partial \varepsilon_{ij}} \right) \Delta \varepsilon_{ij}$$

Equating the above two expressions of dW, we have

$$g_{ij} = \gamma \, \delta_{ij} + \left(\frac{\partial \gamma}{\partial \varepsilon_{ij}} \right) \quad (i, j = 1, 2)$$

This is the relationship between the surface stress and the surface tension. Clearly, the surface tension is only a component of the surface stress tensor. The term $\left(\frac{\partial \gamma}{\partial \varepsilon_{ij}} \right)$ reflects the change of surface tension with respect to the surface strain, i.e., the surface tension dependence on the surface deformation. In other words, generally, surface stress is different from surface tension and surface tension will change with the surface deformation (area change, curvature change, etc.). For most simple liquid-fluid interfaces where the atomic or molecular mobility is high and there is no long range correlation in atomic or molecular positions, the surface tress will be isotropic with zero shear components, i.e., $\left(\frac{\partial \gamma}{\partial \varepsilon_{ij}}\right) = 0$, consequently, the surface stress and the surface tension are numerically equal.

Without loss of generality, in this chapter, we will consider that the solid surfaces are rigid and non-deformable, and their surface free energy is represented by the surface tension.

Sessile Drop on a Flat Horizontal Surface

Consider a sessile drop resting on a solid surface in equilibrium with the liquid's vapor phase (or another fluid phase), as illustrated in the figure below. Please note that the contact angle θ is defined as the angle inside the liquid between the tangent line of the solid–liquid interface and the tangent line of the liquid–vapor interface at the three-phase contact point.

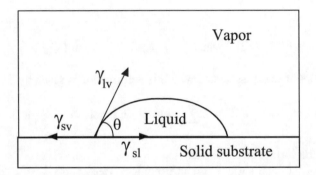

Note that the surface tension is a tensile force, always tends to minimize the surface area (just like the elastic force of a stretched rubber sheet) and hence to minimize the total surface free energy. The direction of a surface tension of an interface in the above figure is from the three-phase contact point towards to its own surface.

As we know, the thermal equilibrium conditions and the chemical equilibrium conditions are the same for both bulk phase systems and surface systems. Our focus of the equilibrium conditions will be on the **mechanical equilibrium conditions**. It should be mentioned here that the entropy maximum principle is the most general principle and can be used in any cases. However, for the purpose of illustrating how to use different thermodynamic approaches, we will derive the mechanical equilibrium conditions by the energy minimum principle for the sessile drop system.

Also, it should be pointed out that we will consider the solid–liquid-vapor three-phase contact line, i.e., the line phase (denoted by the subscript σ in the equations below), in this model and hence its contribution to the mechanical equilibrium condition.

For the sessile drop system illustrated in the figure above, there are a total of 7 phases: three bulk phases (solid, liquid and vapor), three surface phases (solid–vapor interface, solid–liquid interface and liquid–vapor interface), and one line phase (the three-phase contact line). As the bulk solid is rigid (no volume change) and has constant mass, it can be easily proved that the bulk solid phase does not contribute to the equilibrium conditions. Hence we will not include the bulk solid phase in the following thermodynamic model.

The total internal energy of the sessile drop system in differential form is given by:

$$
\begin{aligned}
dU &= dU_L + dU_V + dU_{LV} + dU_{SV} + dU_{SL} + dU_\sigma \\
&= T_L dS_L - P_L dV_L + \sum \mu_{iL} dN_{iL} \\
&\quad + T_V dS_V - P_V dV_V + \sum \mu_{iV} dN_{iV} \\
&\quad + T_{LV} dS_{LV} + \gamma_{LV} dA_{LV} + \sum \mu_{iLV} dN_{iLV} \\
&\quad + T_{SV} dS_{SV} + \gamma_{SV} dA_{SV} + \sum \mu_{iSV} dN_{iSV} \\
&\quad + T_{SL} dS_{SL} + \gamma_{SL} dA_{SL} + \sum \mu_{iSL} dN_{iSL} \\
&\quad + T_\sigma dS_\sigma + \sigma dL + \sum \mu_{i\sigma} dN_{i\sigma}
\end{aligned}
$$

In the above equation, the subscripts L, V, LV, SV, SL and σ stand for the liquid phase, vapor phase, liquid–vapor interface, solid–vapor interface, solid–liquid interface and line phase, respectively.

As our interest here is to find the mechanical equilibrium conditions, we will assume that the thermal and chemical equilibrium conditions are already reached. That is, temperatures of different phases are equal, and chemical potentials of the same component in different phases are also equal. It should be pointed out that these equilibrium conditions are true and use of these conditions will not reduce the generality of our results. In fact, we can carry all the terms in the above equation, use the proper constraint conditions (total entropy constant, total mass constant, etc.), and derive the same equilibrium conditions as below.

By considering the existence of the thermal and chemical equilibrium conditions, the above equation is reduced to

$$
\begin{aligned}
dU &= TdS + \sum \mu_i dN_i - P_L dV_L - P_V dV_V \\
&\quad + \gamma_{LV} dA_{LV} + \gamma_{SV} dA_{SV} + \gamma_{SL} dA_{SL} + \sigma dL
\end{aligned}
$$

where S and N_i are the total entropy and the total mole number of component i, respectively. Note that for a given system, the total internal energy will be minimum at an equilibrium state under the following constraints:

$$
S_{total} = \text{constant} \quad \text{and} \quad N_{i\,total} = \text{constant}
$$

Using these constraints, the above equation becomes:

$$
dU = -P_L dV_L - P_V dV_V + \gamma_{LV} dA_{LV} + \gamma_{SV} dA_{SV} + \gamma_{SL} dA_{SL} + \sigma dL
$$

Use the following geometry constrains:

$$A_{SV} + A_{SL} = \text{constant} \quad \text{and} \quad V_L + V_V = \text{constant}$$

The above equation can be rewritten as:

$$dU = (P_V - P_L)dV_L + \gamma_{LV}dA_{LV} + (\gamma_{SL} - \gamma_{SV})dA_{SL} + \sigma dL$$

It should be noted that the surface area of the liquid–vapor interface of this sessile drop depends on the volume of the drop and the area of the solid–liquid interface (the wetted surface area under the drop); also, the length of the three-phase contact line depends on the area of the solid–liquid interface. That is,

$$A_{LV} = A_{LV}(V_L, A_{SL}) \text{ and } L = L(A_{SL})$$

$$dA_{LV} = \left(\frac{\partial A_{LV}}{\partial V_L}\right)dV_L + \left(\frac{\partial A_{LV}}{\partial A_{SL}}\right)dA_{SL}$$

$$dL = \left(\frac{\partial L}{\partial A_{SL}}\right)dA_{SL}$$

Therefore,

$$dU = \left[-(P_L - P_V) + \gamma_{LV}\left(\frac{\partial A_{LV}}{\partial V_L}\right)\right]dV_L$$
$$+ \left[\gamma_{LV}\left(\frac{\partial A_{LV}}{\partial A_{SL}}\right) - (\gamma_{SV} - \gamma_{SL}) + \sigma\left(\frac{\partial L}{\partial A_{SL}}\right)\right]dA_{SL}$$

Set $dU = 0$, we obtain the following equilibrium conditions:

$$\text{Laplace Equation} \quad (P_L - P_V) = \gamma_{LV}\left(\frac{\partial A_{LV}}{\partial V_L}\right)$$

$$\text{Young's Equation} \quad \gamma_{LV}\left(\frac{\partial A_{LV}}{\partial A_{SL}}\right) + \sigma\left(\frac{\partial L}{\partial A_{SL}}\right) = \gamma_{SV} - \gamma_{SL}$$

The first equation of these two mechanical equilibrium conditions is the Laplace equation, as we discussed before. It is the mechanical equilibrium condition for the liquid-fluid interface, valid for every point at the interface. It governs the equilibrium shape of the liquid-fluid interface.

The second equation in the mechanical equilibrium condition is called the Young's equation.

$$\gamma_{LV}\left(\frac{\partial A_{LV}}{\partial A_{SL}}\right) + \sigma\left(\frac{\partial L}{\partial A_{SL}}\right) = \gamma_{SV} - \gamma_{SL}$$

Let us first evaluate the two derivatives in this equation. If we look at a small region near the three-phase contact line, as illustrated in the figure below, we may consider the profile of the liquid–vapor interface approximately a straight line, as illustrated in the figure below.

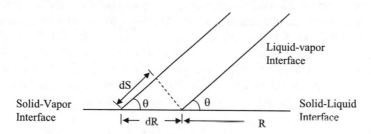

Consider a small variation of the radius of the three-phase contact circle, dR, as shown in the figure above. Assume that the contact angle remains the same during such the small variation, and the corresponding change in the length of the liquid–vapor interface profile is dS. Thus, we have

$$dA_{SL} = 2\pi R dR$$

$$dA_{LV} = 2\pi R dS$$

and

$$dS = dR \cos \theta$$

This yields

$$\left(\frac{\partial A_{LV}}{\partial A_{SL}} \right) = \frac{dS}{dR} = \cos \theta$$

Also, using

$$dL = 2\pi dR \quad \text{and} \quad dA_{SL} = 2\pi R dR,$$

we have

$$\left(\frac{\partial L}{\partial A_{SL}} \right) = \frac{1}{R}$$

Therefore, the Young's equation becomes:

$$\gamma_{LV} \cos \theta = (\gamma_{SV} - \gamma_{SL}) - \frac{\sigma}{R}$$

where R is the radius of the three-phase contact circle. This equation is the equilibrium condition of the three-phase contact line, valid at every point of the three-phase contact line. It implies that the contact angle is a function of the surface tensions, line tensions and the drop size.

It should be noted that the above Young's equation is valid only for a drop on an ideal solid surface. For a sessile drop resting on a non-ideal solid surface, the contact angle will depends not only on the surface tensions and line tensions, but also on surface conditions such as the surface roughness and heterogeneity. In such a case, the three-phase contact line will not be a smooth circle. The Young equation will take more complicated forms depending on surface's conditions. More discussion on this matter is beyond the scope of this chapter and can be found in the literature.

We may rearrange the above Young's equation as follows:

$$\cos\theta = \frac{\gamma_{SV} - \gamma_{SL}}{\gamma_{LV}} - \frac{\sigma}{\gamma_{LV}}\frac{1}{R}$$

When the radius of the three-phase contact circle becomes infinitely large, $R \to \infty$, i.e., an infinite large sessile drop, the above equation reduces to

$$\cos\theta_\infty = \frac{\gamma_{SV} - \gamma_{SL}}{\gamma_{LV}}$$

In the above equation, θ_∞ is the equilibrium contact angle of an infinitely large sessile drop. This equation is referred to as the classical Young equation. It shows that the contact angle of an infinite large sessile drop (or in the case of a straight three-phase contact line) is determined by the three surface tensions of such a solid–liquid-vapor system.

Combining the above two equations yields:

$$\cos\theta = \cos\theta_\infty - \frac{\sigma}{\gamma_{LV}}\frac{1}{R}$$

This equation is referred to as the modified Young's Equation.

For a given solid–liquid-vapor system, the surface tensions are constant, hence θ_∞ or $\cos\theta_\infty$ is a constant. In addition, the line tension σ and the liquid surface tension γ_{LV} are also constant. The modified Young equation implies that $\cos\theta$ is a linear function of the curvature of the three-phase contact circle, $(1/R)$. In other words, the modified Young equation predicts a drop size dependence of the contact angle. If we consider line tension is a positive quantity as the liquid surface tension, the contact angle will decrease as the drop size increases. As shown in the following figure, the slope of the $\cos\theta \sim (1/R)$ line is $\left(\frac{\sigma}{\gamma_{LV}}\right)$. In this way, the line tension can be determined by measuring the drop size dependence of contact angles.

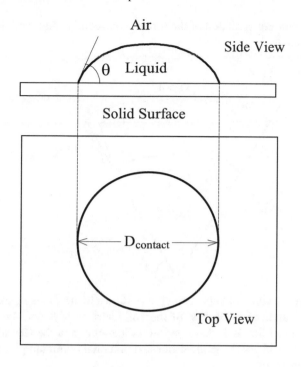

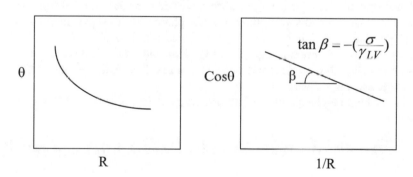

A Sessile Drop in a Crevice/Cavity

The Young's equation derived previously is valid for contact angle formed on a horizontal surface. In the studies of drop or bubble nucleation, the initial drop/bubble nuclei are often found to form first in the crevices or cavities of a substrate. This is because nucleation in such a place requires less energy and hence is easier to form in comparison with the nuclei formed in the homogeneous fluid phase or on the flat surface. More discussions on this aspect will be given in a later section about nucleation. It should be pointed out here that the equilibrium contact angle plays an important role in the nucleation. Therefore, we will show in this section how to derive the equilibrium condition of the three-phase contact line for a drop/bubble in a crevice or a cavity.

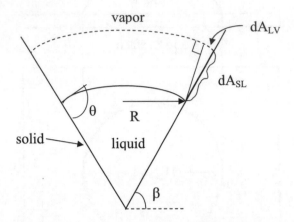

Consider a drop siting in a conic crevice as shown in the above figure. The contact angle is θ, the angle of inclination of the solid surface is β, and the radius of the three-phase contact line is R. Assume the temperature and the chemical potentials are constant. We will use the grand canonical potential minimum principle to derive the equilibrium condition.

For system illustrated in the figure above, we will consider a total of 6 phases: two bulk phases (liquid and vapor), three surface phases (solid–vapor interface, solid–liquid interface and liquid–vapor interface), and one line phase (the three-phase contact line). We will use subscripts L, V, LV, SV, SL and σ to indicate the liquid phase, vapor phase, liquid–vapor interface, solid–vapor interface, solid–liquid interface and the line phase, respectively.

The differential change of the total grand canonical potential of this system is given by

$$d\Omega = -P_L dV_L - P_V dV_V + \gamma_{LV} dA_{LV} + \gamma_{SV} dA_{SV} + \gamma_{SL} dA_{SL} + \sigma dL$$

Using the constraints:

$$V_V + V_L = \text{constant}$$

$$A_{SV} + A_{SL} = \text{constant}$$

We have:

$$d\Omega = -(P_L - P_V)dV_L + \gamma_{LV} dA_{LV} - (\gamma_{SV} - \gamma_{SL})dA_{SL} + \sigma dL$$

Assume the liquid is non-volatile and the liquid volume V_L is constant, and set $d\Omega = 0$ at equilibrium, we have

$$\gamma_{LV} dA_{LV} - (\gamma_{SV} - \gamma_{SL}) dA_{SL} + \sigma dL = 0$$

or

$$\gamma_{LV} \frac{dA_{LV}}{dA_{SL}} - (\gamma_{SV} - \gamma_{SL}) + \sigma \frac{dL}{dA_{SL}} = 0$$

Considering the geometry shown in the above figure and using trigonometry, the following relationships can be derived:

$$\frac{dA_{LV}}{dA_{SL}} = \cos\theta \quad \text{and} \quad \frac{dL}{dA_{SL}} = \frac{\cos\beta}{R}$$

Substituting the above relationships into the equilibrium condition, we obtain the general Young equation for a sessile drop in contact with a conic crevice:

$$\gamma_{LV} \cos\theta = (\gamma_{SV} - \gamma_{SL}) - \frac{\sigma \cos\beta}{R}$$

If the radius of the three-phase contact circle is infinitely large, $R \rightarrow \infty$, the above equation is reduced to:

$$\cos\theta_\infty = \frac{\gamma_{SV} - \gamma_{SL}}{\gamma_{LV}}$$

This leads to the following form of the equilibrium condition:

$$\cos\theta = \cos\theta_\infty - \frac{\sigma}{\gamma_{LV}} \frac{\cos\beta}{R}$$

This equation clearly shows that not only the radius of the three-phase contact line but also the inclination angle of the solid surface affect the equilibrium contact angle. Although derived from a conical crevice system, this equation is valid for a three-phase contact line on any smooth and homogeneous surface of revolution. For example, the three-phase line inside a conical capillary or around a cone.

If the inclination angle is zero, i.e., $\beta = 0$, the crevice becomes a flat surface. Because $\cos\beta = 1$ for this case, the above equilibrium condition becomes the modified Young's equation, as we derived previously:

$$\cos\theta = \cos\theta_\infty - \frac{\sigma}{\gamma_{LV}} \frac{1}{R}$$

If the inclination angle is $\beta = 90°$, the conic crevice becomes a vertical capillary tube of a constant cross-section. In this case, $\cos\beta = 0$, the equilibrium condition becomes

$$cos\theta = cos\theta_\infty$$

It means the line tension and the size of the drop have no influence on the contact angle of a liquid inside a vertical capillary tube.

Home Work

Contact angles of dodecane on FC721 surface are measured as a function of drop size. The results include the following: $\theta_\infty = 70°$, $\gamma_{lv} = 25.4$ mJ/m^2, and

$$\frac{dcos\theta}{d(1/R)} = -7.874 \times 10^{-5} \text{m}.$$

Use the modified Young's Equation to find (a) the line tension value; (b) at what value of the contact radius R the contact angle will approach $\theta = 180°$? what does this mean?

3.6   From Laplace Equation to Capillary Rise and Meniscus Shape

Capillary Rise in a Vertical Capillary

Consider a capillary tube (with a radius R) dipped vertically into a liquid, as illustrated below. The liquid inside the capillary will climb up the tube to a certain height, if the liquid partially wets the inner surface of the capillary tube, i.e., $\theta < 90°$.

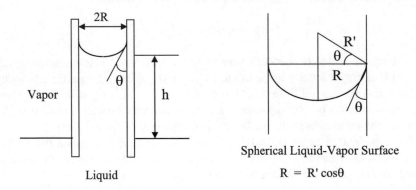

Spherical Liquid-Vapor Surface

R = R' cosθ

If the diameter of the capillary tube is small, we may assume that the liquid–vapor interface is spherical and has a radius R'. According to the Laplace equation, the pressure difference across the spherical liquid–vapor interface is given by:

$$\Delta P = \frac{2\gamma}{R'}$$

Using the geometric relationship between R and R': $R = R'\cos\theta$, we have

$$\Delta P = \frac{2\gamma\cos\theta}{R}$$

Because this pressure difference is responsible to lift the liquid column to a height h against the gravity, we have:

$$\Delta P = \Delta\rho g h$$

where $\Delta\rho$ is the density difference between the liquid and fluid phases. Combining the above two equations gives the capillary height:

$$h = \frac{2\gamma}{\Delta\rho g}\frac{\cos\theta}{R}$$

Example: If we are given $\gamma = 70 \text{ mJ/m}^2$, $\Delta\rho = 1000 \text{ kg/m}^3$, $g = 10 \text{ m/s}^2$, and $\theta = 0°$, we will find

$$h = \frac{1.4\times10^{-5}}{R}\,(\text{m})$$

If the radius of the capillary tube is $R = 1$ mm, the above equation predicts the capillary height is $h = 1.4$ mm. If $R = 1$ μm, it follows $h = 14$ m. This example shows the capillary rise dependence on the capillary size in the extreme case of complete wetting, $\theta = 0°$.

Capillary Rise at a Vertical Wall

In this part, we will see how to determine the meniscus profile and the capillary height by the Laplace equation. Let us consider a solid plate vertically dipped into a liquid. The liquid partially wets the solid surface (i.e., $\theta < 90°$) and hence climbs up and form a capillary rise meniscus, as illustrated in the figure below.

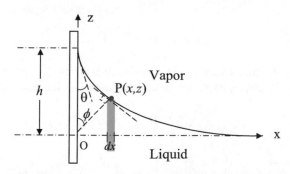

In this figure, the x-axis is at the same level as the horizontal liquid–vapor interface. The origin of the z-axis is also at the horizontal liquid–vapor interface level. θ is the contact angle, h is the height of the capillary rise, and ϕ is the angle between the z-axis and the normal line at an arbitrary point P (x, z). Such a liquid–vapor interface is a cylindrical surface as it curved only in the plane of the paper (x–z plane), therefore, one of the curvatures is zero, i.e., $\frac{1}{R_2} = 0$. The Laplace equation for this surface is

$$\Delta P = \frac{\gamma}{R_1} = \frac{\gamma}{R}$$

Consider a differential liquid column at an arbitrary point P (x, z) on the surface. According to the force balance, the Laplace pressure difference should be equal to the gravity force against lifting that liquid column. That is,

$$\Delta P = \Delta \rho g z$$

and from calculus, we know

$$\frac{1}{R} = -\frac{d \cos \phi}{dz}$$

Thus the Laplace equation becomes:

$$\Delta \rho g z = -\gamma \left(\frac{d \cos \phi}{dz} \right)$$

If we define $a = \frac{\Delta \rho g}{\gamma}$ as a material property of the system, the above equation can be rearranged as:

$$-\frac{d \cos \phi}{dz} = az$$

Integrating the above equation gives:

$$-\cos \phi = \frac{az^2}{2} + b$$

Using the boundary condition: $z = 0$ at $\phi = 0$, it can be shown that $b = -1$. Therefore,

$$\cos \phi = -\frac{az^2}{2} + 1$$

This equation describes the meniscus profile in terms of the variables z and ϕ. Using another boundary condition: $z = h$ at $\phi = 90°–\theta$, the above equation gives:

$$h = \sqrt{\frac{2\gamma}{\Delta \rho g}} \sqrt{1 - \sin \theta} = c \sqrt{1 - \sin \theta}$$

where $c = \sqrt{\frac{2\gamma}{\Delta \rho g}}$ is usually referred to as the capillary constant. The capillary constant value indicates the capability of a liquid to climb up on a vertical wall when $\theta = 0°$ (i.e., complete wetting), or the maximum height of the capillary rise. For example, at 20 °C, if the liquid is water, the vapor phase is the water vapor or air, the liquid–vapor surface tension is approximately 73 mJ/m². The capillary constant has a value: $c = 3.86$ mm. At 20 °C, if the liquid is a soap solution, the vapor phase is air, the liquid–air surface tension is approximately 32 mJ/m². In this case, the capillary constant has a value: $c = 2.56$ mm.

This equation relates the height of the capillary rise to material properties of the fluids (i.e., the capillary constant) and the contact angle. We can rearrange the above equation as:

$$\sin \theta = 1 - \frac{h^2}{c^2}$$

This is a very useful equation. It allows us to determine the contact angle by measuring the height of the capillary rise at a vertical plate. Since the height (i.e., length) can be measured very accurately, the contact angle determined in this way is also very accurate. The well-known Wilhelmy Plate method for contact angle measurement is based on this equation and has an accuracy of 0.1 degree.

Capillary Rise and the Meniscus Shape with an Inclined Plate

Let us consider a tilted solid plate partially inserted in a liquid, as shown in the figure below. Assume that both surfaces of the plate are the same and are partially wetted by the liquid. The contact angle of the liquid on both sides of the solid plate is θ. The liquid surface tension is γ, and the angle of inclination of the plate is β. The heights of capillary rise on the two sides of the solid plate are different and are represented by h_L on the left-hand side and h_R on the right-hand side, respectively.

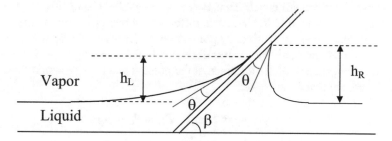

For this system, we want to find (1) the equilibrium shape or the profile of the liquid–vapor interfaces on both sides of the plate; and (2) the heights of the capillary rise on both sides of the plate.

As we know, the equilibrium shape of any liquid-fluid interface is governed by the Laplace equation of capillarity:

$$\Delta P = \gamma \left(\frac{1}{R_1} + \frac{1}{R_2} \right)$$

Since the liquid–vapor interfaces in this system are cylindrical surfaces,

$$\tfrac{1}{R_2} = \tfrac{1}{\infty} = 0$$

the Laplace equation becomes

$$P_V - P_L = \gamma \tfrac{1}{R_1} = \tfrac{\gamma}{R} \tag{3.6.1}$$

At this point, in order to find the profile of the liquid–vapor interface, we need to find how ΔP is related to $1/R$ in terms of variables that describe the profile of the liquid–vapor interface. Let us consider the liquid–vapor interface on the left side of the plate, as illustrated below.

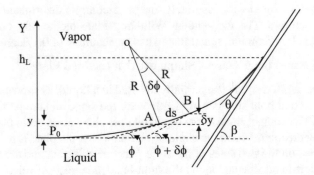

First, we know that the bulk fluid phase pressures on both sides of the horizontal liquid–vapor interface are equal. That is, $P_L = P_V = P_0$. Using this pressure P_0 as the reference, let us find the pressure difference at any point across the curved liquid–vapor interface. At an arbitrary point A on the interface, the height above the level of the horizontal liquid–vapor interface is y. In the above figure, ϕ is the angle between the tangent line of the interface at the point A and the horizontal line.

Therefore, at point A, we have:

$$P_V = P_0 - \rho_v g y \quad \text{and} \quad P_L = P_0 - \rho_L g y$$

Thus

$$P_V - P_L = (\rho_L - \rho_v) g y \tag{3.6.2}$$

Knowing the expression of the pressure difference in the Laplace equation at point A, we need to evaluate the curvature $(1/R)$ at point A, before we can find information about the profile of the liquid–vapor interface. From the elementary calculus, we can show (see the above figure):

$$\tfrac{1}{R} = \tfrac{\delta \phi}{\delta S}$$

and

$$\delta y = \delta S \sin \phi$$

Combining the above two relations gives:

$$\frac{1}{R} = \sin \psi \frac{\delta \phi}{\delta y} \tag{3.6.3}$$

Bringing the above expressions for ΔP and $(1/R)$ into the Laplace equation yields:

$$\Delta \rho g y = \gamma \sin \phi \frac{\delta \phi}{\delta y}$$

or

$$\Delta \rho g y \delta y = \gamma \sin \phi \delta \phi$$

Integrating this equation with the boundary condition: $\phi - 0$ at $y = 0$, we have:

$$y = 2 \left(\frac{\gamma}{\Delta \rho g} \right)^{1/2} \sin \frac{\phi}{2} \tag{3.6.4}$$

Equation (3.6.4) describes the equilibrium shape of the liquid–vapor interface on the left side of the plate.

For the liquid–vapor interface on the right hand side of the plate, using the same procedure and realizing the angle ϕ is negative, we can show the profile of the liquid–vapor interface on the right side of the plate is given by the same equation, Eq. (3.6.4).

Now let us evaluate the capillary heights of the left-side and right-side liquid–vapor interfaces. The capillary height on the left side can be obtained by applying the following boundary condition to Eq. (3.6.4):

$$y = h_L, \quad \phi = \beta - \theta$$

$$h_L = \sqrt{2} C \sin \frac{\beta - \theta}{2} \tag{3.6.5}$$

where the capillary constant C is defined as:

$$C = \sqrt{\frac{2\gamma}{\Delta \rho g}}.$$

Similarly, applying the boundary condition of the right side

$$y = h_R, \quad \phi = \pi - \beta - \theta$$

to Eq. (3.6.4) yields:

$$h_R = \sqrt{2} C \cos \tfrac{\beta+\theta}{2} \tag{3.6.6}$$

Comparing Eq. (3.6.5) with Eq. (3.6.6), we see generally

$$h_R \neq h_L$$

$$\frac{h_L}{h_R} = \frac{\sin\left(\frac{\beta-\theta}{2}\right)}{\cos\left(\frac{\beta+\theta}{2}\right)}$$

For example, $\beta = 70°$, $\theta = 30°$, $\frac{h_L}{h_R} = 0.53 < 1$. The capillary height on the right side of the plate is higher than that on the left side.

Now let us consider an extreme case where $\beta = 0°$. For example, as illustrated in the figure below, we may hang a plate horizontally on the liquid surface and then pull the plate up to a height h_0 above the undisturbed liquid level.

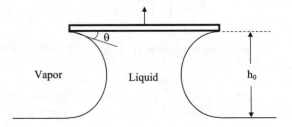

For this case, Eq. (3.6.6) gives:

$$h_0 = \sqrt{2} C \cos \tfrac{\theta}{2} \tag{3.6.7}$$

For a given liquid–vapor system, if the contact angle θ is specified, the above equation will give the maximum height of the liquid column. If one lifts the plate any further, i.e., $h > h_0$, the liquid will detach from the plate, or the meniscus will break. For example, consider water at $T = 20 °C$, from Eq. (3.6.7) we have:

$\theta = 0°$	$h_0 = 5.40$ mm
$\theta = 30°$	$h_0 = 5.22$ mm
$\theta = 60°$	$h_0 = 4.67$ mm
$\theta = 90°$	$h_0 = 3.81$ mm
$\theta = 120°$	$h_0 = 2.70$ mm
$\theta = 180°$	$h_0 = 0.00$ mm

A Liquid Bridge between Two Parallel Plates

From our daily life experience, we know that a drop of water placed between two clean and smooth glass plates may draw the two plates together, if the contact angle $\theta < 90°$. One may need to apply a considerable force in order to separate them. However, if one places a non-wetting liquid ($\theta > 90°$) between two closely spaced plates, then a huge force may be required to push the two plates any closer to each other.

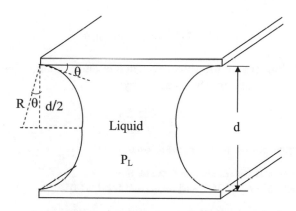

As shown in the above figure, the contact angle between the liquid and the solid surface is θ, and the separation distance between the two plates is d. Let us define:

A_S as the solid–liquid contact area on each plate.

L as the length of the solid–liquid-vapor three-phase contact line on each plate.

The force acting to pull these two plates together is in the direction perpendicular to the parallel plates and given by:

$$F = (P_V - P_L)A_S + (\gamma \sin \theta)L$$

As we can see from this equation, the force F has two components:

(1) $(P_V - P_L)A_S$ is due to the Laplace pressure, caused by the surface tension and the curvature of the liquid–vapor interface.
(2) $(\gamma \sin \theta)L$ is the component of the surface tension in the direction perpendicular to each plate.

Consider the plates are infinitely long in the direction perpendicular to the paper plane. The meniscus can then be assumed to be cylindrical, i.e., $(1/R_2) = 0$. The Laplace equation in this case becomes

$$P_V - P_L = \gamma \left(\frac{1}{R_1} \right) = \frac{\gamma}{R}$$

Because the separation distance between the two plates is small, we can assume that the shape of the liquid–vapor interface in the plane of the paper is spherical. By inspecting the above figure, we have:

$$R \cos \theta = \frac{d}{2} \quad \text{or} \quad R = \frac{d}{2 \cos \theta}$$

Therefore,

$$\Delta P = \frac{2\gamma \cos \theta}{d}$$

and

$$F = \frac{2\gamma \cos \theta}{d} A_S + \gamma L \sin \theta$$

Usually, the 2nd term in the above equation is much smaller than the 1st term. Therefore, the force is approximately given by:

$$F = \frac{2\gamma \cos \theta}{d} A_S$$

From this equation, we may conclude that:

(1) F increases as d decrease. $F \to \infty$ as $d \to 0$.
(2) F is attractive for $\theta < 90°$, and repulsive for $\theta > 90°$.

Example: The strong repulsive force between two surfaces is used to lubricate moving metallic elements, for example, bearings. We can choose an oil as a lubricant (liquid) and treat the metal surface so that $\theta > 90°$, thus the metal surfaces are repulsed and hence preventing them from sliding over each other directly causing excessive friction wear. Let us assume that the contact angle is 120°. The force per unit area can be evaluated using the above equation as follows:

$\theta = 120°$	$\gamma = 30$ mJ/m^2	$d = 1.0$ μm	$F/A_S = -0.3$ atm
$\theta = 120°$	$\gamma = 30$ mJ/m^2	$d = 0.1$ μm	$F/A_S = -3.0$ atm
$\theta = 120°$	$\gamma = 50$ mJ/m^2	$d = 0.1$ μm	$F/A_S = -5.0$ atm

In the above table, the negative sign of the force indicates that the force is repulsive. Clearly, the repulsive force increases significantly as the surface tension increases and as the separation distance decreases.

3.7 Curvature Effects on Equilibrium Pressure and Temperature

Curvature Effect on Equilibrium Vapor Pressure Around a Drop

Consider a liquid drop in equilibrium with its vapor phase at a constant temperature.

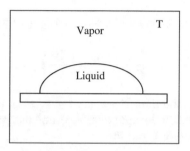

We know that at an equilibrium state, the chemical potentials of the liquid and the vapor phases must be the same. That is,

$$\mu_L = \mu_V \quad \text{or} \quad d\mu_L = d\mu_V \tag{3.7.1}$$

Under the condition that T = constant,

$$d\mu = -sdT + vdP = vdP.$$

Therefore, the chemical equilibrium condition becomes

$$v_L dP_L = v_V dP_V \tag{3.7.2}$$

Assume the drop surface is spherical. From the Laplace equation

$$P_L - P_V = \frac{2\gamma}{R}$$

we have:

$$dP_L = dP_V + d\left(\frac{2\gamma}{R}\right) \tag{3.7.3}$$

Using Eq. (3.7.3) to replace the dP_L in Eq. (3.7.2), we have:

$$v_L dP_V + v_L d\left(\frac{2\gamma}{R}\right) = v_V dP_V$$

$$dP_V(v_L - v_V) = -v_L d\left(\frac{2\gamma}{R}\right)$$

Because $v_L << v_V$,

$$dP_V = \frac{v_L}{v_V} d\left(\frac{2\gamma}{R}\right) \qquad (3.7.4)$$

Assume the vapor as an ideal gas: $P_V v_V = \bar{R}T$, Eq. (3.7.4) can be rewritten as:

$$\frac{dP_V}{P_V} = \frac{v_L}{\bar{R}T} d\left(\frac{2\gamma}{R}\right)$$

Using the boundary condition: $R \to \infty$, $P_V = P_{V\infty}$ ($P_{V\infty}$ is the equilibrium vapor pressure over a flat liquid–vapor interface), and integrating the above equation gives the well-known **Kelvin Equation**:

$$\ln \frac{P_V}{P_{V\infty}} = \frac{v_L}{\bar{R}T}\left(\frac{2\gamma}{R}\right) > 0 \qquad (3.7.5)$$

The Kelvin equation provides a relationship between the equilibrium vapor pressure over a drop and the radius of the drop. As the right-hand side of Eq. (3.7.5) is positive, it implies that

$$P_V > P_{V\infty}$$

That is, a convex liquid–vapor interface (e.g., the drop surface) requires a higher equilibrium vapor pressure than a flat liquid–vapor surface at the same temperature. However, the curvature effect on the vapor pressure usually is small, i.e., $\frac{P_V}{P_{V\infty}} \approx 1$, unless the radius of the drop is sufficiently small, e.g., $R < 1\ \mu$m. In the case of drop nucleation, the initial size of a nucleus is very small, thus it requires very high vapor pressure to form a stable nucleus. That is why drop nucleation (e.g., dropwise condensation) is possible only in a supersaturated vapor environment.

Finally, it is worthwhile to point out that Eq. (3.7.4) is derived by using the chemical equilibrium condition and the mechanical equilibrium condition (the Laplace equation). The drop radius in this equation should be the radius required by a drop in equilibrium with the vapor phase at the given temperature. Therefore, the radius in the Kelvin equation is also called the **equilibrium radius**, R_e, of a drop for a given vapor pressure P_V. Rearranging Eq. (3.7.5) gives:

$$R_e = \frac{v_L}{\bar{R}T} \frac{2\gamma}{\ln \frac{P_V}{P_{V\infty}}} = f\left(T, \gamma, \frac{P_V}{P_{V\infty}}\right) \qquad (3.7.6)$$

From this equation, it is obvious that the equilibrium radius of the droplet is proportional to the surface tension of the liquid–vapor interface, γ. If the liquid is water, its surface tension is approximately 72.8 mJ/m² at 20 °C. It is well-known that adding surfactant to water can reduce its surface tension significantly, for example, close to zero at high concentration of surfactant. Therefore, one can expect that there are only very small droplets in the presence of high concentration of surfactant in water.

Curvature Effect on Equilibrium Pressure in a Bubble

Consider a bubble surrounded by its liquid phase. Using a similar approach to the derivation of the Kelvin equation, we can show:

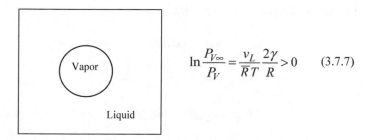

$$\ln \frac{P_{V\infty}}{P_V} = \frac{v_L}{\overline{R}T} \frac{2\gamma}{R} > 0 \qquad (3.7.7)$$

Equation (3.7.7) implies that $P_V < P_{V\infty}$, i.e., the equilibrium vapor pressure inside a bubble is lower than that over a flat liquid–vapor interface at the same temperature. In other words, a concave liquid–vapor surface requires a lower equilibrium vapor pressure than a flat liquid–vapor surface at the same temperature. This is the cause of the "capillary condensation" phenomenon. At a given temperature, the partial vapor pressure in the air may be below the saturation pressure, and hence no moisture condensation from air would be expected. However, vapor condensation may be observed in some porous materials such as an insulation layer or soil. This is because the menisci in the fine capillaries or crevices may be concave surfaces and have very small radii of curvatures. According to Eq. (3.7.7), the smaller the R, the smaller the ratio $(P_V/P_{V\infty})$. Therefore, these menisci require much lower equilibrium vapor pressure, and hence the air near these menisci is saturated. Condensation will thus occur in these fine capillaries. Such a phenomenon is called capillary condensation.

Similarly, the equilibrium radius of the bubble is given by:

$$R_e = \frac{2v_L}{\overline{R}T} \frac{\gamma}{\ln\frac{P_{V\infty}}{P_V}}$$

This equation indicates that the equilibrium radius of the bubble is proportional to the surface tension of the liquid–vapor interface, γ. As discussed previously for droplets, if the surrounding liquid is water, and surfactant is added to water, the surface tension of the water–vapor interface can be reduced significantly. Therefore, one can expect that there are only very small sized bubbles in the presence of high concentration of surfactant in water.

Curvature Effect on Equilibrium Temperature

Consider a droplet in a vapor phase inside a piston-cylinder arrangement. The system is in contact with a pressure reservoir so that the pressure of the vapor phase is constant, $P_V =$ constant. From the Gibbs–Duhem equation,

$$d\mu_V = -s_V dT + v_V dP_V = -s_V dT \tag{3.7.8}$$

$$d\mu_L = -s_L dT + v_L dP_L \tag{3.7.9}$$

Using the chemical equilibrium condition:

$$d\mu_L = d\mu_V \tag{3.7.10}$$

we have:

$$(s_V - s_L)dT = -v_L dP_L \tag{3.7.11}$$

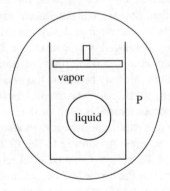

Note that

$$(s_V - s_L) = \frac{\Delta h_{LV}}{T}$$

where Δh_{LV} is the latent heat of evaporation, and

$$dP_L = d\left(\frac{2\gamma}{R}\right)$$

(from the Laplace equation), Eq. (3.7.11) becomes:

$$\frac{\Delta h_{LV}}{T} dT = -v_L d\left(\frac{2\gamma}{R}\right) \tag{3.7.12}$$

Assume Δh_{LV}, v_L and γ are constant with respect to the curvature ($1/R$), integrating Eq. (3.7.12) from $R = \infty$ to a finite value of R yields:

$$\ln \frac{T}{T_\infty} = -\frac{v_L}{\Delta h_{LV}} \frac{2\gamma}{R} < 0 \tag{3.7.13}$$

This equation is called **Thomson Equation** (derived by Thomson in 1888). T_∞ in this equation is the equilibrium temperature (i.e., phase equilibrium temperature) of a bulk liquid phase with a flat liquid-vapor interface ($R \to \infty$). Equation (3.7.11) indicates that $T < T_\infty$. This means that, at a given vapor pressure, the equilibrium temperature required for a droplet is lower than that required for a bulk liquid with a flat liquid-vapor interface. In other words, at a given vapor pressure, it is easier for a droplet to evaporate as it requires a lower equilibrium temperature or boiling temperature. On the other hand, in order to have a vapor phase condensed into small droplets, it is necessary to cool the vapor below the saturation temperature T_∞ of the bulk liquid at the same pressure.

Similarly, for a bubble in a liquid phase with $P_L = $ constant, we can show

$$\left(\frac{1}{T_\infty} - \frac{1}{T}\right) = \frac{R_u}{\Delta h_{LV}} \ln\left[\frac{2\gamma}{RP_L} + 1\right] > 0 \tag{3.7.14}$$

where R_u is the universal gas constant. Equation (3.7.14) indicates that the equilibrium temperature of the vapor inside a bubble is higher than the equilibrium temperature of the vapor over a flat liquid-vapor interface, i.e., $T(R) > T_\infty$. That means, it is necessary to raise the temperature above T_∞ for a small vapor bubble to exist in a saturated liquid under a given saturation pressure $P_L(T_\infty)$. In other words, it requires a superheated liquid to generate small bubbles. From Eq. (3.7.14), we see that the smaller the bubble, the greater the superheating (i.e., $T - T_\infty$) is required. This conclusion is consistent with what we found about the curvature effect on equilibrium vapor pressure inside a bubble. As an example, let us consider pure water at 1 atm. Using Eq. (3.7.14) we can find that the equilibrium temperature for a bubble of 1 μm in radius is 126 °C (compare this with $T_\infty = 100°C$ for bulk liquid water with a flat liquid-vapor interface)!

Home Work

(1)   Consider a single component system that consists of a spherical liquid droplet surrounded by vapor. The vapor and droplet are contained in a piston-cylinder device. The liquid phase may be treated as incompressible and the vapor phase is treated as an ideal gas.

(a) If the cylinder is surrounded by a thermal reservoir, which function
behaves as the thermodynamic potential for the system? What conditions must the intensive properties satisfy in order for equilibrium to
exist?

(b) Show the equilibrium radius of the drop is:

$$R_e = 2\gamma_{lv}/[P_\infty - P_v + (R_u T/v_L) \ln(P_v/P_\infty)]$$

where P_∞ is the equilibrium pressure of a bulk liquid or vapor phase at a given
temperature without surface effect; R_u is the universal gas constant.

(c) If the temperature of the reservoir is 100 °C, show a sketch of R_e versus P_v,
if the liquid and vapor are water. The surface tension γ_{lv} is about 60 mJ/m^2 at
100 °C.

(2) A spherical bubble of a liquid vapor is in equilibrium with a pure liquid phase
(liquid and the vapor are the same molecules) contained in a piston-cylinder
arrangement. The cylinder is in contact with a pressure reservoir and can freely
exchange thermal energy with the surrounding. Assume vapor in the bubble
behaves as an ideal gas.

(a) Find the boiling point (the equilibrium temperature) as a function of the
radius of the bubble, R. You may set $P_\infty = P_L$.

(b) Consider the liquid is water and the liquid pressure $P_L = 1$ atm, the boiling
point in the case of a planar liquid–vapor interface T $(R = \infty) = 373$ K,
the latent heat $L = 40.4$ kJ/kmol, surface tension $\gamma = 55.46 - 0.215$
(T – 373) mJ/m^2, find the boiling point value at $R = 1$ μm.

3.8 Solute Effect on Equilibrium Radius of Droplets

Consider a drop in equilibrium with its vapor phase at a constant temperature and
constant pressure. The liquid of the drop is a dilute solution of two components:
the solvent (component 1) and the solute (component 2 and nonvolatile). The mole
fraction of the solute is x. Because the solute is not volatile, the vapor phase contains
only component1 (the solvent).

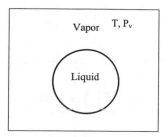

The chemical potential of the solvent is given by:

$$\mu_{1L}(T, P, x) = \mu_{1L}^0(T, P) - xR_uT$$

$$d\mu_{1L} = d\mu_{1L}^0(T, P_L) - d(xR_uT) = -s_LdT + v_LdP_L - d(R_uTx) \qquad (3.8.1)$$

In the above equations, in order to distinguish from the drop radius R, we use R_u to indicate the universal gas constant.

Assume the vapor phase contains only the component 1.

$$d\mu_{1V} = -s_VdT + v_VdP_V \qquad (3.8.2)$$

We know that at an equilibrium state, the chemical potentials of the liquid and the vapor phases must be the same. That is,

$$\mu_L = \mu_V \quad \text{or} \quad d\mu_L = d\mu_V \qquad (3.8.3)$$

Under the condition that $T = $ constant, and $P_V = $ constant,

$$d\mu_{1L} = v_LdP_L - R_uTdx$$

$$d\mu_V = 0$$

Therefore, the chemical equilibrium condition becomes

$$v_LdP_L - RTdx = 0 \qquad (3.8.4)$$

Assume that the drop surface is spherical. From the Laplace equation

$$P_L - P_V = \frac{2\gamma}{R} \qquad (3.8.5)$$

we have:

$$dP_L = dP_V + d\left(\frac{2\gamma}{R}\right) \tag{3.8.6}$$

Using Eq. (3.8.6) to replace the dP_L in Eq. (3.8.4), and noting $dP_V = 0$, we have:

$$v_L d\left(\frac{2\gamma}{R}\right) = R_u T dx$$

$$\frac{dR}{dx} = -\frac{R_u T}{v_L}\left(\frac{R^2}{2\gamma}\right) \tag{3.8.7}$$

This equation indicates that the equilibrium radius of the droplet will decrease with the increase of the solute molar fraction. In other words, under the same conditions (T and P_V), the equilibrium size of a solution droplet is smaller than the equilibrium size of a pure water droplet.

Next, we will explore the possibility of finding an expression of the equilibrium radius of the droplet as an explicit function of the mole fraction. The chemical potentials of the solvent in the vapor phase and the liquid phase can be written as follows:

$$\mu_v(T, P_v) = \mu_v(T, P_\infty) + R_u T \ln\left(\frac{P_v}{P_\infty}\right) \tag{3.8.8}$$

$$\begin{aligned}\mu_L(T, P_L, x) &= \mu_{1L}^0(T, P_L) - R_u T x \\ &= [\mu_L(T, P_\infty) + v_L(P_L - P_\infty)] - R_u T x\end{aligned} \tag{3.8.9}$$

In these equations, P_∞ is the equilibrium pressure or the saturation vapor pressure over a flat liquid–vapor interface at the given temperature T. P_V is the equilibrium vapor pressure in the system with a liquid drop. P_L is the pressure inside the drop. x is the mole fraction.

At equilibrium,

$$\mu_l(T, P_l, x) = \mu_v(T, P_v) \tag{3.8.10}$$

and

$$\mu_L(T, P_\infty) = \mu_V(T, P_\infty).$$

From Eqs. (3.8.8) and (3.8.9), we have

$$v_L(P_L - P_\infty) - R_u T x = R_u T \ln\left(\frac{P_v}{P_\infty}\right) \tag{3.8.11}$$

From Laplace equation, we have

$$P_L = P_V + \frac{2\gamma}{R}$$

Equation (3.8.11) becomes:

$$v_L\left(P_V - P_\infty + \frac{2\gamma}{R}\right) - R_u Tx = R_u Tln\left(\frac{P_V}{P_\infty}\right) \qquad (3.8.12)$$

If the drop radius is not too small, e.g., $R > 1$ μm, the curvature effect on the vapor pressure usually is very small, i.e., $\frac{P_v}{P_\infty} \approx 1$. In such a case, Eq. (3.8.12) is reduced to:

$$v_L\left(\frac{2\gamma}{R}\right) = R_u Tx$$

or

$$R = \frac{2\gamma v_L}{R_u Tx}$$

Clearly, as the mole fraction increases, the equilibrium radius of the droplet will decrease. In other words, the equilibrium radius of the droplet is smaller if the mole fraction of the solute in the liquid is higher.

3.9 Heterogeneous Nucleation

When a new thermodynamic phase is formed, nucleation is the first step. For example, when a vapor is cooled to a temperature below its saturation temperature, nucleuses in the form of tiny liquid droplets may form as the beginning of condensation process, i.e., a new liquid phase is forming from the vapor. If liquid water is heated to a temperature above its saturation temperature, nucleuses in the form of tiny bubbles may form as the liquid water starts boiling or vaporizing, i.e., a new vapor phase is forming.

Generally, there are two types of nucleation: homogeneous nucleation and heterogeneous nucleation. In homogeneous nucleation, nucleus forms in a uniform bulk phase. For example, minute liquid droplets form in a vapor phase. In heterogeneous nucleation, nucleus forms on a surface. For example, dew droplets form on leaves of grass. As it will be demonstrated at the end of this section, the surface will reduce the energy required to form a nucleus, and hence make the heterogeneous nucleation easier than the homogeneous nucleation.

Consider the formation of a liquid droplet on a solid surface in contact with a super-saturated vapor phase. In the initial state, the system has only a vapor phase in contact with a solid surface. In the final state, a liquid droplet nucleus is formed on the solid surface. The process is illustrated in the figure below. We assume the system is sufficiently large, forming one small droplet will not affect the temperature and vapor pressure of the system, i.e., $T_1 = T_2 = T$, and $P_{1V} = P_{2V} = P_V$.

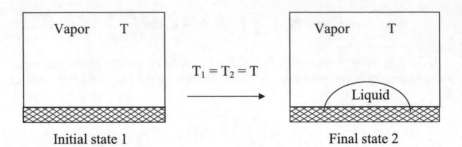

Initial state 1   Final state 2

Since this is an isothermal process, we use the Helmholtz potential to model the system. Recall:

$$F_B = U - TS = -PV + \sum \mu_i N_i \quad \text{for a bulk phase}$$

$$F_A = U - TS = \gamma A + \sum \mu_i N_i \quad \text{for a surface phase}$$

At state 1, the system has a vapor phase and a solid–vapor interface. The total free energy of the system is:

$$F_{1,total} = \sum F_{B1} + \sum F_{A1} = -P_{1v}V_{1v} + N_{1v}\mu_{1v} + \gamma_{sv}A_{sv1} \tag{3.9.1}$$

where the subscripts B and A stand for bulk phase and surface phase, respectively; subscript 1 stands for the initial state 1, and subscripts v and sv stands for the vapor phase and the solid–vapor interface, respectively.

At state 2, the system has a vapor phase, a liquid phase, a solid–vapor interface, a solid–liquid interface and a liquid–vapor interface. The total free energy of the system at state 2 is:

$$
\begin{aligned}
F_{2,total} &= \sum F_{B2} + \sum F_{A2} \\
&= [-P_{2v}V_{2v} + N_{2v}\mu_{2v}] + [-P_{2l}V_{2l} + N_{2l}\mu_{2l}] \\
&\quad + \gamma_{lv}A_{lv} + \gamma_{sv}A_{sv2} + \gamma_{sl}A_{sl}
\end{aligned}
\tag{3.9.2}
$$

where the subscript 2 stands for the final state 2, the subscript l represents the liquid phase, the subscripts lv and sl represent the liquid–vapor interface and the solid–liquid interface, respectively; and the subscripts v and sv stands for the vapor phase and the solid–vapor interface, respectively.

In the above, we have assumed that the number of molecules in the surface phases is negligible in comparison with that in the bulk phases. Therefore, all $(\mu_i N_i)$ terms for surface phases are neglected.

Because $T_1 = T_2 = T$, and $P_{1V} = P_{2V} = P_V$, it follows that

$$\mu_{1v}(T, P_{1v}) = \mu_{2v}(T, P_{2v}) = \mu_v$$

From Eqs. (3.9.1) and (3.9.2), the free energy change of the nucleation process is given by:

$$\begin{aligned}
\Delta F &= F_{2,total} - F_{1,total} \\
&= -(P_{2l} - P_v)V_{2l} + \gamma_{lv}A_{lv} - (\gamma_{sv} - \gamma_{sl})A_{sl} + (\mu_{2l} - \mu_v)N_{2l}
\end{aligned} \tag{3.9.3}$$

where the following constraints were used:

$$V_{1v} = V_{2v} + V_{2l}$$

$$A_{sv1} = A_{sv2} + A_{sl}$$

$$N_{1v} = N_{2v} + N_{2l}$$

$$\mu_{1v} = \mu_{2v} = \mu_v$$

If we assume that the tiny nucleus is a spherical cap on the solid surface, we have the following geometrical relationships.

$$P_l - P_v = \frac{2\gamma_{lv}}{R}$$

$$\gamma_{lv}\cos\theta = \gamma_{sv} - \gamma_{sl}$$

$$A_{lv} = 2\pi R^2(1 - \cos\theta)$$

$$A_{sl} = \pi R^2 \sin\theta$$

$$V_l = \frac{\pi R^3}{3}(1 - \cos\theta)^2(2 + \cos\theta)$$

In Eq. (3.9.3), $(P_{2l} - P_v)$, $(\gamma_{sv} - \gamma_{sl})$, V_{2l}, A_{lv} and A_{sl} can be evaluated via the above geometrical relations. The attention should be given to the chemical potential term $(\mu_{2l} - \mu_v)$. From Chap. 1, we know that the chemical potentials of the vapor phase and the liquid phase can be written as follows:

$$\mu_v(T, P_v) = \mu_v(T, P_\infty) + R_u T \ln\left(\frac{P_v}{P_\infty}\right)$$

$$\mu_{2l}(T, P_{2l}) = \mu_l(T, P_\infty) + v_l(P_{2l} - P_\infty)$$

In these equations, P_∞ is the equilibrium pressure or the saturation vapor pressure over a flat liquid–vapor interface at the given temperature T. P_v is the super-saturated vapor pressure in the system. R_u indicates the universal gas constant. When the liquid phase and the vapor phase at equilibrium,

$$\mu_L(T, P_\infty) = \mu_V(T, P_\infty)$$

Therefore,

$$(\mu_{2l} - \mu_v) = v_l(P_{2l} - P_\infty) - R_u T \ln\left(\frac{P_v}{P_\infty}\right) \tag{3.9.4}$$

Putting Eq. (3.9.4) into Eq. (3.9.3), we have:

$$\Delta F = -(P_{2l} - P_v)V_{2l} + \gamma_{lv}A_{lv} - (\gamma_{sv} - \gamma_{sl})A_{sl}$$
$$+ N_{2l}v_l(P_{2l} - P_\infty) - N_{2l}R_u T \ln\left(\frac{P_v}{P_\infty}\right)$$

It should be realized that $V_{2l} = N_{2l}\,v_l$. Thus, the above equation of free energy change of the nucleation process can be re-written as

$$\Delta F = \underbrace{(P_v - P_\infty)V_{2l} - N_{2l}R_uT \ln\left(\frac{P_v}{P_\infty}\right)}_{\text{related to the bulk liquid phase}} + \underbrace{\gamma_{lv}A_{lv} - (\gamma_{sv} - \gamma_{sl})A_{sl}}_{\text{related to the surface phases}}$$

$$= \Delta F_B \quad + \quad \Delta F_A \tag{3.9.5}$$

As we see, the first two terms on the right-hand side of this equation are related to the newly formed bulk liquid phase (i.e., V_{2l} and N_{2l}). Let us call these two terms the free energy change of the bulk phase, ΔF_B. The last two terms on the right-hand side of this equation are related to the three surfaces: the solid–vapor interface, and the newly formed liquid–vapor interface and the solid–liquid interface. Let us call these two terms the free energy change of the surface phases, ΔF_A.

Let's compare the magnitude of the two terms in ΔF_B. If we assume that the liquid is water at a room temperature (20 °C), and the nucleus' size is $R = 100$ nm, using the Kelvin equation yields:

$$\frac{(P_v - P_\infty)V_{2l}}{N_{2l}R_uT \ln\left(\frac{P_v}{P_\infty}\right)} \cong \frac{7}{1000}$$

Thus, $(P_v - P_\infty)V_{2l}$ in Eq. (3.9.5) can be neglected for reasonably large nuclei (i.e., R > 10 nm). The free energy of nucleation is reduced to

$$\Delta F = \underbrace{-N_{2l}R_uT \ln\left(\frac{P_v}{P_\infty}\right)}_{\Delta F_B} + \underbrace{\gamma_{lv}A_{lv} - (\gamma_{sv} - \gamma_{sl})A_{sl}}_{\Delta F_A} \tag{3.9.6}$$

In the above equation, ΔF is the total free energy change required creating a single nucleus. If ΔF is negative, it means that forming the nucleus will reduce the free energy of the system and the system is moving towards an equilibrium state with minimum energy. Such a nucleation process is thermodynamically favorable and can occur spontaneously. However, if ΔF is positive, it means that forming the nucleus requires increasing the free energy of the system. Increasing the system's energy is possible only by additional energy input from outside. Such a process cannot occur naturally or spontaneously. The larger the positive ΔF, the more difficult to form a nucleus.

From Eq. (3.9.6), one can see that the free energy change associated with the creation of a new bulk liquid phase, ΔF_B, **is a negative value**, and contributes to reduce the total free energy of the system. However, the free energy change associated with the creation of the new surfaces, ΔF_A, **is positive**, and hence contributes to increase the total free energy of the system. This will become more evident by the next equation, Eq. (3.9.7), as shown below. These competing effects clearly depend on the size of the nucleus.

Use the geometrical relations for a spherical cap and $V_{2l} = N_{2l}\, v_l$, as well as the Young's equation, Eq. (3.9.6) becomes:

$$\Delta F = -\frac{\pi R^3}{3\overline{v}_l}(1 - \cos\theta)^2(2 + \cos\theta)R_u T \ln\left(\frac{P_v}{P_\infty}\right)$$
$$+ \pi R^2 \gamma_{lv}(1 - \cos\theta)^2(2 + \cos\theta) \qquad (3.9.7)$$

In this equation, R is the radius of the spherical nucleus, R_u is the universal gas constant, $\overline{v}_l$ is the specific molar volume of the liquid, and θ is the contact angle. Clearly, the first term in Eq. (3.9.7) is the free energy change caused by the formation of the new bulk liquid phase, ΔF_B, and is a negative quantity. The second term is the free energy change required by the formation of the surface phases, ΔF_A, and is a positive quantity. The relative magnitudes of these two terms determine if the total free energy change of the nucleation process is positive or negative.

From this equation, Eq. (3.9.7), we know that for a given solid–liquid-vapor system at a given condition (T, P$_v$), the surface tension γ_{lv} and the contact angle θ are constant, therefore, the total free energy change ΔF will depend on the nucleus size, R. If we plot ΔF vs R, we will have a figure as illustrated below.

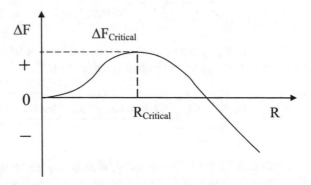

As seen from the figure above, initially, as the radius of the nucleus increases, the free energy change of nucleation is positive and increases too. This means that a larger amount of energy is required to create a nucleus. According to Eq. (3.9.6) or Eq. (3.9.7), this is because the free energy change is required by the formation of the new surface phases, and ΔF_A is dominant. Gradually, the total free energy of nucleation reaches a maximum value, called the critical free energy of nucleation, $\Delta F_{critical}$. The corresponding radius of the nucleus is called the critical radius of nucleus, $R_{critical}$. When the radius of the nucleus is larger than the critical radius, the free energy change of nucleation decreases with the increase of the radius of the nucleus. According to Eq. (3.9.6) or Eq. (3.9.7), this is because the free energy change due to the formation of the new bulk liquid phase, ΔF_B, becomes dominant. Especially after the radius of the nucleus is larger than a certain value, we can see from the curve in the above figure, the free energy change of nucleation becomes negative. This indicates that the formation of such a nucleus will reduce the total free energy of the system. In other words, such a nucleation is thermodynamically favorable.

Let us find the critical radius of a nucleus at a given condition (T, P_v). Because

$$\frac{\partial(\Delta F)}{\partial R} = 0$$

From Eq. (3.9.7), the above condition results in the critical radius as given by the equation below:

$$R_C = \frac{2\gamma_{lv}\bar{v}_l}{R_u T \ln(P_v/P_\infty)} \tag{3.9.8}$$

It should be noted that Eq. (3.9.8) is just the Kelvin equation we derived in a previous section from the chemical potential equilibrium condition, i.e.,

$$\ln(P_v/P_\infty) = \frac{2\gamma_{lv}\bar{v}_l}{\overline{RT}R_e}$$

Using Eq. (3.9.8), if we replace $R_u T \ln(P_v/P_\infty)$ in Eq. (3.9.7) by $\frac{2\gamma_{lv}\bar{v}_l}{R_C}$, we will have the critical free energy of nucleation as:

$$\Delta F(R_c) = \frac{4\pi}{3}R_C^2 \gamma_{lv} f(\theta) \tag{3.9.9}$$

where

$$f(\theta) = (1 - \cos\theta)^2(2 + \cos\theta)/4$$

$$0 \le f(\theta) \le 1$$

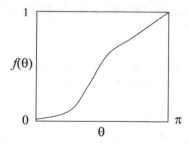

In the above, the function $f(\theta)$ is referred to as the **wetting function**. $f(\theta)$ reflects the substrate effect, i.e., the contact angle effect on the nucleation. Obviously, a smaller value of $f(\theta)$ will reduce the critical free energy of nucleation, and hence corresponds to a favorable condition for the nucleation. In the extreme cases, if $\theta = \pi$, $f(\theta) = 1$. This means that a completely non-wettable surface has no effect (to reduce the ΔF) on the nucleation. Since the substrate surface has no effect on the nucleation, the nucleation becomes the same as the homogeneous nucleation. In fact, if we set $\theta = \pi = 180°$, the previously derived Eq. (3.9.7) becomes the free energy change of homogeneous nucleation. If $\theta = 0$, $f(\theta) = 0$. This means $\Delta F = 0$, or there is no activation energy required for the nucleation on a surface that can be completely wetted by the liquid. In such a case, vapor super-saturation will not occur, because the vapor will spontaneously condense to form a liquid film on the solid surface.

Generally, for a given solid surface, the contact angle increases with the liquid surface tension. For a given liquid, liquid surface tension is fixed, if the solid–vapor surface tension (γ_{sv}) decreases, the contact angle increases, i.e., the surface becomes less wettable. From surface thermodynamics point of view, by changing the surface free energy or surface tension of the solid substrate, for example, by using different materials or by using different coating, the contact angle can be changed. Therefore, the nucleation can be either enhanced or impaired. For instance, lowering the solid surface tension is an important mechanism for anti-fog and anti-frost windows or eye glasses. In the case of boiling heat transfer, lowering the liquid surface tension γ_{lv} by adding some surfactants and hence lowering the contact angle can enhance the nucleation heat transfer.

Home Work

Consider the effects of the three-phase contact line on the nucleation of a liquid droplet on a planar solid surface from a vapor phase.

(1) Derive the free energy of nucleation, ΔF_{hetero}.
(2) Find the critical nucleation radius R_C and the critical free energy of nucleation, ΔF_C.
(3) For pure water at $T = 295$ K, $\gamma = 72$ mJ/m^2, $\sigma = 0.1$ µJ/m, $\theta_\alpha = \pi/4$, plot ΔF vs R for $P_V/P_{V\alpha} = 1.1$ and 1.3.

3.10 Equilibrium Condition of a Bubble in a Uniform Electric Field

Consider a spherical bubble of vapor is surrounded by an aqueous solution, as illustrated in the figure below. The system is subject to a uniform electric field and a constant temperature.

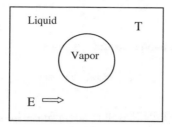

This system consists of two bulk phases: the liquid and the vapor, and one interface phase: the liquid–vapor interface. If the bulk liquid phase is infinitely large in comparison with the size of the bubble, we may consider the liquid phase as a mass reservoir. Therefore, in additional to the constant temperature, the chemical potential of this system is also constant. Therefore we can use the grand canonical free energy to model this system.

If there is no applied electric field, the total grad canonical free energy of this system is given by:

$$\Omega = \Omega_L + \Omega_V + \Omega_{LV} = -(PV)_L - (PV)_V + \gamma_{LV} A_{LV}$$

In the differential form,

$$d\Omega = d\Omega_L + d\Omega_V + d\Omega_{LV} = -P_L dV_L - P_V dV_V + \gamma_{LV} dA_{LV}$$

However, in this case we have an applied electric field. The electric field will induce surface charge on the surface of the bubble. Let us denote the applied electric potential as ΔV, the total surface charge as Q, the surface charge density per unit surface area as σ_{lv} and the surface area of the bubble as A_{lv}. The work done by the applied electrical field to create surface charge on the bubble surface is given by:

$$W_e = \Delta V Q = \Delta V \sigma_{lv} A_{lv}$$

Or, on per unit surface area basis,

$$\frac{W_e}{A_{lv}} = \Delta V \sigma_{lv}$$

Therefore, the total grand canonical free energy should include the work done by the electric field, as:

$$d\Omega = d\Omega_L + d\Omega_V + d\Omega_{LV} + dW_e$$
$$= -P_L dV_L - P_V dV_V + \gamma_{LV} dA_{LV} + \Delta V \sigma_{lv} dA_{LV}$$

At equilibrium state, according to the grand canonical free energy minimum principle,

$$d\Omega = 0$$

The constraints for this bubble–liquid system are:

$$V_L + V_V = constant \quad \text{or} \quad dV_V = -dV_L$$

and the surface area of the bubble is dependent of the bubble's volume,

$$A_{LV} = f(V_V) \quad \text{or} \quad dA_{LV} = \frac{\partial A_{LV}}{\partial V_V} dV_V$$

Using these constraints yields:

$$d\Omega = -(P_V - P_L)dV_V + \gamma_{LV}\left(\frac{\partial A_{LV}}{\partial V_V}\right)dV_V + \Delta V \sigma_{lv}\left(\frac{\partial A_{LV}}{\partial V_V}\right)dV_V$$

$$d\Omega = -(P_V - P_L)dV_V + (\gamma_{LV} + \Delta V \sigma_{lv})\left(\frac{\partial A_{LV}}{\partial V_V}\right)dV_V$$

From

$$d\Omega = 0,$$

we have

$$(P_V - P_L) = (\gamma_{LV} + \Delta V \sigma_{lv})\left(\frac{\partial A_{LV}}{\partial V_V}\right)$$

For a spherical bubble, it can be shown that

$$\left(\frac{\partial A_{LV}}{\partial V_V}\right) = \frac{2}{R}$$

where R is the radius of the spherical bubble.

Finally, the mechanical equilibrium condition of such a spherical bubble in a liquid phase and under an applied electric field is:

$$(P_V - P_L) = \frac{2}{R}(\gamma_{LV} + \Delta V \sigma_{lv})$$

In comparison with the Laplace equation of a bubble without the applied electric field:

$$(P_V - P_L) = \frac{2}{R}\gamma_{LV}$$

we see clearly that, because

$$\Delta V \sigma_{lv} > 0,$$

the effect of the applied electric field is to increase the pressure difference $(P_V - P_L)$. $(\gamma_{LV} + \Delta V \sigma_{lv})$ may be considered as an apparent surface tension, or **electro-surface tension**, that is,

$$\widetilde{\gamma_{LV}} = (\gamma_{LV} + \Delta V \sigma_{lv}).$$

Although the equilibrium condition

$$(P_V - P_L) = \frac{2}{R}(\gamma_{LV} + \Delta V \sigma_{lv})$$

was derived for a spherical bubble suspended in an aqueous solution, it can be shown that, similar to the derivation demonstrated in Sect. 3.5, this equation is valid as the mechanical equilibrium condition for the liquid–vapor interface of a spherical-cap shaped bubble attached on a solid surface. Of course, the mechanical equilibrium condition involving the contact angle at the three-phase contact line will be more complicated when the electric field effects on the solid–vapor surface and the solid–liquid interface are considered.

Furthermore, we know that at an equilibrium state, the chemical potentials of the liquid and the vapor phases must be the same. That is,

$$\mu_L = \mu_V \quad \text{or} \quad d\mu_L = d\mu_V$$

Under the condition of constant temperature, i.e., $T = $ constant,

$$d\mu = -sdT + vdP = vdP.$$

Therefore, the chemical equilibrium condition becomes

$$v_L dP_L = v_V dP_V$$

From the Laplace equation derived above for a bubble in a liquid phase and under an applied electric field:

$$P_V - P_L = \frac{2}{R}(\gamma_{lv} + \Delta V \sigma_{lv}) = \frac{2}{R}\widetilde{\gamma}_{lv}$$

Using the same procedure as shown in Sect. 3.7, we can demonstrate

$$ln \frac{P_{V\infty}}{P_V} = \frac{2v_L}{R_u T} \frac{\tilde{\gamma}_{lv}}{R}$$

where $P_{V\infty}$ is the equilibrium vapor pressure over a flat liquid–vapor interface; v_L is the mole specific volume of the liquid; and R_u is the universal gas constant.

Rearranging the above equation leads to the equilibrium radius of the bubble under the electric field:

$$R_e = \frac{2v_L}{R_u T} \frac{\tilde{\gamma}_{lv}}{ln \frac{P_{V\infty}}{P_V}} = \frac{2v_L}{R_u T} \frac{(\gamma_{lv} + \Delta V \sigma_{lv})}{ln \frac{P_{V\infty}}{P_V}}$$

Clearly, this equation indicates that the equilibrium radius of the bubble will increase with the applied electrical voltage, ΔV.

3.11 Effects of Applied Electrical Field on Contact Angles

(Electro-Wetting Phenomenon)

So far, we know that the shape of a liquid drop on a solid surface, as indicated by the contact angle, depends on the liquid properties (reflected by γ_{lv}), solid surface properties (reflected by $\gamma_{sv,}$ and γ_{sl}). When an electric potential is applied across a liquid drop and a dielectric solid substrate, ions and dipoles will redistribute either in the liquid, or in the solid or both, depending on the relative material properties. This redistribution of charges can change the contact angle. This phenomenon, i.e. change in the contact angle of a droplet on a solid surface by applying an electric field between a conducting liquid and a solid substrate is called "electro-wetting".

Figure below shows the schematic of an electro-wetting phenomenon. As seen from the figure, there is a dielectric substrate between the electrode and the liquid. The dielectric substrate between the electrode and the liquid is used to block the electron transfer and prevent chemical oxidation and electrolysis. Additionally, the dielectric substrate provides a hydrophobic surface that usually has a large initial contact angle.

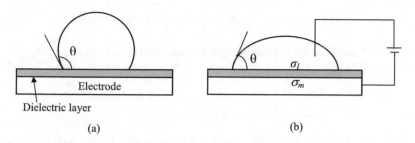

Schematic of electro-wetting phenomenon (a) No electric field. (b) With applied electric field

When an electric field is applied as illustrated in the figure above, the electric charge at the liquid–solid interface changes, which results in the change of total free energy of the system and the spreading of the droplet over the solid surface. By applying a voltage difference cross the liquid and the electrode, there is a charge density σ_l in the liquid phase at the solid–liquid interface, and an induced charge density σ_m on the metal electrode. It should be noted that, across the dielectric layer, there are surface charge on the solid–liquid interface and induced charge on the surface of the metal electrode. This configuration acts like a parallel plate capacitor. Meanwhile, the contact angle changes and consequently causes an increase in the solid–liquid interfacial area (dA). This increase causes a change in the total free energy of the system in terms of surface free energies (surface tensions) and additional electric energies required to create the charge density in the liquid and on the electrode (σ_l and σ_m).

Based on the above analysis, the differential form of the total grand canonical free energy of the system can be written as:

$$d\Omega - \gamma_{sl}dA_{sl} + \gamma_{sv}dA_{sv} + \gamma_{lv}dA_{sl}cos\theta + d\Omega_{de} - dW_e$$

where γ_{sl}, γ_{sv}, and γ_{lv} are the surface tension of the solid–liquid, the solid–vapor and the liquid–vapor interfaces; A_{sl} and A_{sv} are the surface area of the solid–liquid and the solid–vapor interfaces; θ is the contact angle; Ω_{de} is the energy stored in the dielectric layer between liquid and electrode, W_e is the electrical field work to create surface charge at the solid–liquid interface by the applied voltage.

It must be mentioned that, when deriving the Young's equation in Sect. 3.5, we have a figure illustrating the geometric relations at the three-phase contact line. Using that figure, we have shown that

$$dA_{lv} = dA_{sl}cos\theta.$$

This relation can be used here.

Since $A_{sl} + A_{sv} = $ constant, $dA_{sl} = -dA_{sv} = dA$, the above equation can be written as:

$$d\Omega = \gamma_{sl}dA - \gamma_{sv}dA + \gamma_{lv}dA\cos\theta + d\Omega_{de} - dW_e \qquad (3.11.1)$$

Let's first consider the case of no applied electric filed, i.e. $d\Omega_{de} = dW_e = 0$. To find the equilibrium conditions, free energy of the system should be minimized, i.e.

$$\frac{d\Omega}{dA} = 0.$$

Applying this principle to the above equation results in Young's equation, as expected:

$$\gamma_{lv}\cos\theta = \gamma_{sv} - \gamma_{sl}$$

Now, consider the case that an electric field is applied to the system. The electrostatic energy in the dielectric layer can be considered as the electrostatic energy stored in a parallel plate capacitor, and is given by:

$$\Omega_{de} = \tfrac{1}{2}CV^2 = \tfrac{1}{2}\left(\tfrac{Q}{V}\right)V^2 = \frac{1}{2}QV = \frac{1}{2}V(\sigma_l A)$$

where V is the electrical potential difference across the dielectric layer; σ_l is the surface charge density on the liquid side of the dielectric layer; A is the solid surface area covered by the drop, i.e., the area of the solid–liquid interface. On per unit surface area basis,

$$\frac{\Omega_{de}}{A} = \frac{1}{2}V\sigma_l$$

Therefore, the change in electrostatic energy per unit area in the dielectric layer upon increase of droplet base (dA) is given by:

$$\frac{d\Omega_{de}}{dA} = \frac{1}{2}V\sigma_l \qquad (3.11.2)$$

The work done by the applied electrical field to create surface charge (Q) at the solid–liquid interface (droplet base) is given by:

$$W_B = VQ = V\sigma_l A$$

or

$$\frac{dW_B}{dA} = V\sigma_l \qquad (3.11.3)$$

Substituting Eqs. (3.11.3) and (3.11.2) into Eq. (3.11.1) gives:

$$d\Omega = \gamma_{sl}dA - \gamma_{sv}dA + \gamma_{lv}dA\cos\theta_{ew} + \left(\frac{1}{2}V\sigma_l - V\sigma_l\right)dA$$

$$d\Omega = \gamma_{sl}dA - \gamma_{sv}dA + \gamma_{lv}dA\cos\theta_{ew} - \frac{1}{2}V\sigma_l dA$$

where θ_{ew} is the contact angle in the presence of applied electric field.
To find the equilibrium conditions, let us set $\frac{d\Omega}{dA} = 0$. This gives:

$$\gamma_{lv}\cos\theta_{ew} = \gamma_{sv} - \gamma_{sl} + \frac{1}{2}V\sigma_l \qquad (3.11.4)$$

As we consider the dielectric layer as a parallel plate capacitor, and the thickness of the dielectric layer is δ, the surface charge density is given by:

$$\sigma_l = \varepsilon_0\varepsilon_r\frac{V}{\delta}.$$

where ε_0 is the dielectric permittivity in vacuum and ε_r is the relative dielectric constant of the liquid.
Substituting this relation back into Eq. (3.11.4) gives:

$$\cos\theta_{ew} = \left(\frac{\gamma_{sv} - \gamma_{sl}}{\gamma_{lv}} + \frac{\varepsilon_0\varepsilon_r V^2}{2\gamma_{lv}\delta}\right) = \left(\cos\theta + \frac{\varepsilon_0\varepsilon_r V^2}{2\gamma_{lv}\delta}\right) \qquad (3.11.5)$$

In the above equation, θ is the contact angle without the applied electric field (given by the Young's equation), and θ_{ew} is the contact angle with the applied electric field. Because the second term on the right-hand side of Eq. (3.11.5) is a positive quantity, Eq. (3.11.5) implies $\theta_{ew} < \theta$.

The above equation indicates that an applied electric field can cause a decrease in the contact angle value. This means that the applied electric field tends to make the drop to spread on the solid surface. As the applied electrical potential increases, the contact angle decreases more and more. However, the complete wetting of the surface by the drop under applied electric field has never been seen in the experimental studies. This implies the equation derived above may be valid only under low applied voltage. At high electric field, many other mechanisms such as electrolysis may have to be considered in the model.

Home Work

Plot θ_{ew} as a function of the applied voltage V, given the equilibrium contact angle θ is 120 degrees, liquid surface tension is 72 mJ/m^2, the dielectric film thickness is 10 μm, the dielectric constant is 4.

3.12 Effects of Electric Double Layer on Contact Angle

When a charged surface in contact with an electrolyte solution, the electrostatic charges on the surface will attract the counter-ions and repel the co-ions in the liquid. This will result in more counter-ions than co-ions in a thin region of the liquid near the charged surface. The net charge in this thin liquid region is to balance the charge on the surface. The region with the net charge of counter-ions and the charged surface are call the **electric double layer**. Generally, the solid–liquid interfaces for all solid surfaces in contact with aqueous electrolyte solutions (including pure water) have electric double layer. Interfaces between aqueous electrolyte solutions and their vapor phases or air have electric double layer on the liquid side.

Recall that surface tension is the free energy required to create a unit area of a surface or interface, additional electric energy is required to create electric double layer at a surface. The electrostatic interaction energy between the charges on the surface and the counter-ions in the liquid side of electric double layer is determined by the electric potential of the charged surface and the total net counter-ion charge. Because the net counter-ion charge in the liquid side of electric double layer is to balance the charge on the surface, the total net counter-ion charge is equal to the total charge on the surface. The electric interaction energy for a liquid–vapor interface is given by

$$\Omega_{EDL-lv} = \psi_{lv}\sigma_{lv}A_{lv}.$$

The electric interaction energy for a solid–liquid interface is given by

$$\Omega_{EDL-sl} = \psi_{sl}\sigma_{sl}A_{sl}$$

In these equations, ψ is the electrostatic potential of the interface, σ is the surface charge density of the interface, A is the surface area.

Consider a sessile drop resting on a solid surface in equilibrium with the liquid's vapor phase (or another fluid phase), as illustrated in the following figure. For simplicity, we assumed the solid–vapor surface does not have any surface charge. However, both the solid–liquid interface and the liquid–vapor interface have surface charges and an electric double layer. From thermodynamic point of view, these surfaces have additional electric energy. How will the electric double layers of the solid–liquid interface and the liquid–vapor interface influence the equilibrium contact angle?

As we have demonstrated previously, the bulk phases (solid, liquid and vapor) will not contribute to the equilibrium condition at the three-phase contact line, therefore, we will consider only the surface phases in the following thermodynamic model.

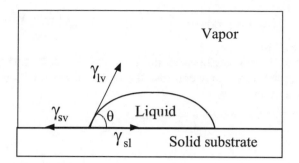

The differential form of the total grand canonical free energy change of the system can be written as:

$$d\Omega = \gamma_{sl}dA_{sl} + \gamma_{sv}dA_{sv} + \gamma_{lv}dA_{lv} + d\Omega_{EDL-lv} + d\Omega_{EDL-sl}$$

where γ_{lv}, γ_{sv} and γ_{sl} are interfacial tensions of the liquid–vapor interface, the solid–vapor interface and the solid–liquid interface, respectively; Ω_{EDL-lv} is the free energy associated with creating the electric double layer of the liquid–vapor surface, Ω_{EDL-sl} is the free energy associated with creating the electric double layer of the solid–liquid interface.

$$d\Omega = \gamma_{sl}dA_{sl} + \gamma_{sv}dA_{sv} + \gamma_{lv}dA_{lv} + \psi_{lv}\sigma_{lv}dA_{lv} + \psi_{sl}\sigma_{sl}dA_{sl}$$

Use geometry constraint:

$$A_{SV} + A_{SL} = \text{constant}$$

we have:

$$d\Omega = \left[(\gamma_{sl} + \psi_{sl}\sigma_{sl}) - \gamma_{sv}\right]dA_{sl} + (\gamma_{lv} + \psi_{lv}\sigma_{lv})dA_{lv}$$

At the equilibrium state, the total free energy of this system should be minimum, therefore,

$$d\Omega = 0.$$

This leads to

$$\left[(\gamma_{sl} + \psi_{sl}\sigma_{sv}) - \gamma_{sv}\right]dA_{sl} + (\gamma_{lv} + \psi_{lv}\sigma_{lv})dA_{lv} = 0$$

or

$$\left[(\gamma_{sl} + \psi_{sl}\sigma_{sl}) - \gamma_{sv}\right] = -(\gamma_{lv} + \psi_{lv}\sigma_{lv})\left(\frac{dA_{lv}}{dA_{sl}}\right)$$

If we look at a small region near the three-phase contact line, as illustrated in the following figure, we may consider the profile of the liquid–vapor interface approximately a straight line.

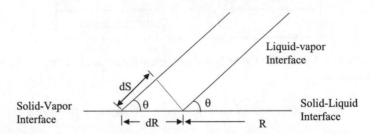

Consider a small variation of the radius of the three-phase contact circle, dR. Assume that the contact angle remains the same during such a small variation, and the corresponding change in the length of the liquid–vapor interface profile is dS. Thus, we have

$$dA_{SL} = 2\pi R dR$$

$$dA_{LV} = 2\pi R dS$$

and

$$dS = dR \cos\theta$$

This yields

$$\left(\frac{\partial A_{LV}}{\partial A_{SL}}\right) = \frac{dS}{dR} = \cos\theta$$

The equilibrium condition becomes:

$$\left[(\gamma_{sl} + \psi_{sl}\sigma_{sl}) - \gamma_{sv}\right] = -(\gamma_{lv} + \psi_{lv}\sigma_{lv})\cos\theta$$

If we approximate the electrostatic potential of the interface ψ by the zeta potential of the electric double layer, ζ, the above equation becomes:

$$\left[(\gamma_{sl} + \zeta_{sl}\sigma_{sl}) - \gamma_{sv}\right] = -(\gamma_{lv} + \zeta_{lv}\sigma_{lv})\cos\theta$$

For a flat surface, the surface charge density is given by

$$\sigma = \frac{4n_\infty ze}{k} \sinh\left(\frac{ze\zeta}{2k_bT}\right)$$

where $k = \left(2n_oz^2e^2/\varepsilon\varepsilon_ok_bT\right)^{1/2}$ is the Debye–Huckel parameter, n_∞ and z are the bulk ionic concentration and the valence of ions, respectively, e is the charge of a proton, $\varepsilon\varepsilon_0$ is the dielectric constant of the liquid, κ_b is the Boltzmann constant, and T is the absolute temperature. ζ is the zeta potential of the electric double layer.

Assuming weak zeta potentials and using the linear approximation (i.e., $\sinh(x) \approx x$, if $x < 1$)), we have

$$\sigma = \frac{2n_\infty z^2 e^2}{k_bTk}\zeta$$

Finally, we have

$$\left[\left(\gamma_{sl} + \frac{2n_\infty z^2 e^2}{k_bTk}\zeta_{sl}^2\right) - \gamma_{sv}\right] = -\left(\gamma_{lv} + \frac{2n_\infty z^2 e^2}{k_bTk}\zeta_{lv}^2\right)\cos\theta$$

$$\frac{2n_\infty z^2 e^2}{k_bTk}\left(\zeta_{sl}^2 + \zeta_{lv}^2\cos\theta\right) = (-\gamma_{lv}\cos\theta + \gamma_{sv} - \gamma_{sl})$$

If we assume the surface tensions, γ_{lv}, γ_{sv} and γ_{sl}, are independent of the surface charge,

$$\gamma_{LV}\cos\theta_0 = (\gamma_{SV} - \gamma_{SL})$$

where θ_0 is the equilibrium contact angle without considering the electric double layer effects.

$$\frac{2n_\infty z^2 e^2}{k_bT\,k}\left(\zeta_{sl}^2 + \zeta_{lv}^2\cos\theta\right) = \gamma_{lv}[\cos\theta_0 - \cos\theta]$$

Generally, the values of the zeta potentials ζ_{sl} and ζ_{lv} are different; however, the difference is not very large. In order to analyze the above equation qualitatively, we assume that ζ_{sl} and ζ_{lv} have the same value, ζ. Under this assumption, the above equation can be written as:

$$A\zeta^2(1 + \cos\theta) = \gamma_{lv}(\cos\theta_0 - \cos\theta)$$

where

$$A = \frac{2n_\infty(ze)^2}{k_BTK}$$

The above equation can be re-arranged as:

$$\frac{(\cos\theta_0 - \cos\theta)}{(1 + \cos\theta)} = \frac{A}{\gamma_{lv}}\zeta^2 > 0$$

If θ is less than 90°, $\cos\theta$ is positive and decreases with θ. The above equation implies that $\theta > \theta_0$. That is, the presence of the zeta potential increases the contact angle. The larger the zeta potential, the larger the increase of the contact angle.

3.13   Modelling Surface Processes by Using Surface Free Energy

Surface Free Energy

Consider a column of a material with a cross-section area A in a fluid (gas or liquid). If we do work to it by cutting it into two parts and separating these two parts sufficiently far away from each other, two new surface areas, 2A, are created (see the figure below). The work done to create these new surfaces is proportional to the created surface area, that is,

$$W = \gamma(2A) \quad (J)$$

The proportional coefficient γ is the surface tension. From this equation, we have

$$\gamma = W/2A \quad (J/m^2)$$

That is why the surface tension is also called the free energy per unit surface area. In addition, because the surface tension and the surface area are all positive quantities, this work is a positive quantity. In other words, in order to create the new surfaces, external work must be done.

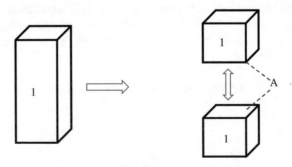

Cohesion

When the above process is reversed, that is, two columns of the same material are brought into contact and form one column, two surfaces (surface area 2A) are eliminated, as shown in the figure below. The free energy change of this process is:

$$\Omega_2 - \Omega_1 = 0 - \gamma(2A) = -2\gamma A < 0$$

Because the free energy change of this process is negative, i.e., the system reduces its free energy, such a process can occur spontaneously, without external work input.

The bonding of the same molecules or the alike-molecules is called the cohesion. The free energy of cohesion is therefore given by:

$$\Delta\Omega_{cohesion} = -2\gamma A < 0$$

We can expect that two particles of the same material or two cells of the same type in a fluid phase will always tend to attract and bind to each other.

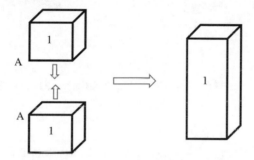

Adhesion

The bonding of different molecules or the dislike-molecules is called the adhesion. Consider two columns of different materials in a fluid phase. If the two columns bind to each other, two original surfaces are eliminated and a new interface is formed (see the figure below). The free energy change of this process is given by:

$$\Delta\Omega_{adhesion} = \Omega_2 - \Omega_1 = \gamma_{12}A_{12} - (\gamma_1 A_1 + \gamma_2 A_2)$$

This free energy change may be positive or negative, depending on the interfacial tension γ_{12}, surface tensions γ_1 and γ_2, the new interfacial surface area A_{12}, and the two original surface areas, A_1 and A_2. In other words, whether these two columns attach to each other or not, it depends on the values of these parameters.

If we assume $A_1 = A_2 = A_{12} = A$, the above equation can be simplified as

$$\Delta\Omega_{adhesion} = (\gamma_{12} - \gamma_1 - \gamma_2)A$$

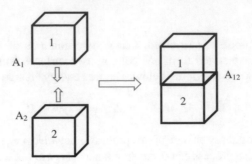

This principle can be applied to model the engulfment of a particle or droplet by another droplet, or the engulfment of one cell by another cell.

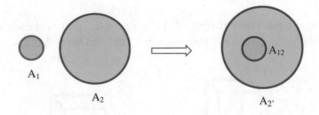

$$\Delta\Omega_{\text{engulf}} = \Omega_2 - \Omega_1 = \gamma_{12}A_{12} + \gamma_2 A_{2'} - (\gamma_1 A_1 + \gamma_2 A_2)$$

or

$$\Delta\Omega_{\text{engulf}} = \gamma_{12}A_{12} + \gamma_2(A_{2'} - A_2) - \gamma_1 A_1$$

If the free energy change is negative, such an engulfment is thermodynamically favorable, because the system always moves towards the minimization of its total free energy.

Below are two examples of applying the free energy change to model interfacial processes.

Merge of Two Oil Droplets in Water

Consider two oil droplets suspended in water under a constant temperature. When they approach to each other, they merge and form one larger droplet, as shown in the figure below.

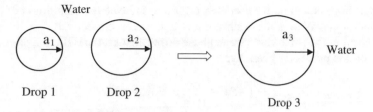

Water

Drop 1   Drop 2

Drop 3

Water

Let us use Helmholtz free energy to model this process.
For bulk phases,

$$F_{bulk} = -PV + \sum \mu_i N_i$$

For surface or interface phases,

$$F_{surf} = \gamma A + \sum (\mu_i N_i)_A$$

In comparison with the mole numbers of the bulk phases, we may assume the mole numbers of a surface phase is negligible:

$$N_{iA} \approx 0$$

Thus,

$$F_{surf} = \gamma A$$

Let us define state 1 as the initial state when the Drop 1 and Drop 2 are not in contact, and state 2 as the final state after the Drop 1 and Drop 2 are merged and Drop 3 is formed. At Sate 1, the system has two oil droplet bulk phases (Drop 1 and Drop 2), two oil–water interface phases and the bulk water phase.

$$F_{state-1} = F_1 + F_{1A} + F_2 + F_{2A} + F_W$$
$$= (-PV + \sum \mu_i N_i)_1 + \gamma_1 A_1 + (-PV + \sum \mu_i N_i)_2$$
$$+ \gamma_2 A_2 + (-PV + \sum \mu_i N_i)_W$$

In the above, we used subscripts 1 and 2 to indicate the bulk liquid of drop 1 and the bulk liquid of drop 2, respectively; subscripts A and W to indicate surface and bulk water, respectively.

At Sate 2, the system has one oil droplet bulk phase (Drop 3), one oil–water interface phase and the bulk water phase.

$$F_{state-2} = F_3 + F_{3A} + F_W$$
$$= (-PV + \sum \mu_i N_i)_3 + \gamma_3 A_3 + (-PV + \sum \mu_i N_i)_W$$

In the above equation, we used subscript 3 to indicate the bulk liquid of drop 3; subscripts A and W to indicate surface and bulk water, respectively.

As the properties of the bulk water phase do not change, the Helmholtz free energy change of this process is given by:

$$F_{state-2} - F_{state-1} = (-PV + \sum \mu_i N_i)_3 + \gamma_3 A_3$$
$$- \left[(-PV + \sum \mu_i N_i)_1 + \gamma_1 A_1 + \left(-PV + \sum \mu_i N_i \right)_2 + \gamma_2 A_2 \right]$$

In order to simplify the analysis, we assume that the two oil drops consist of one and the same molecule. Therefore the above equation can be reduced to:

$$
\begin{aligned}
F_{state-2} - F_{state-1} =& (-P_3 V_3 + \mu_3 N_3) + \gamma_3 A_3 \\
& - \left[(-P_1 V_1 + \mu_1 N_1) + \gamma_1 A_1 + (-P_2 V_2 + \mu_2 N_2) + \gamma_2 A_2 \right] \\
=& [P_1 V_1 + P_2 V_2 - P_3 V_3] + \left[\gamma_3 A_3 - \gamma_1 A_1 - \gamma_2 A_2 \right] \\
& + [\mu_3 N_3 - \mu_1 N_1 - \mu_2 N_2]
\end{aligned}
$$

Let us analyze the three [~] terms in the above equation.

$$
\begin{aligned}
[P_1 V_1 + P_2 V_2 - P_3 V_3] &= P_1 V_1 + P_2 V_2 - P_3 (V_1 + V_2) \\
&= V_1 (P_1 - P_3) + V_2 (P_2 - P_3)
\end{aligned}
$$

Here the volume constrain $V_1 + V_2 = V_3$ is used.

$$[\mu_3 N_3 - \mu_1 N_1 - \mu_2 N_2] = N_1 (\mu_3 - \mu_1) + N_2 (\mu_3 - \mu_2)$$

Here the mass constrain $N_1 + N_2 = N_3$ is used.
Recall the chemical potential of a pure liquid is given by:

$$\mu(T, P) = \mu(T, P_\infty) + v(P - P_\infty)$$

$$v = \frac{V}{N}$$

and assume the specific volume of the oil phase remains constant, $v_1 = v_2 = v_3$, we can show

$$
\begin{aligned}
[\mu_3 N_3 - \mu_1 N_1 - \mu_2 N_2] &= N_1 (\mu_3 - \mu_1) + N_2 (\mu_3 - \mu_2) \\
&= -V_1 (P_1 - P_3) - V_2 (P_2 - P_3)
\end{aligned}
$$

Clearly, from the above analysis,

$$[P_1 V_1 + P_2 V_2 - P_3 V_3] + [\mu_3 N_3 - \mu_1 N_1 - \mu_2 N_2] = 0$$

Consequently,

$$F_{state-2} - F_{state-1} = \gamma_3 A_3 - \gamma_1 A_1 - \gamma_2 A_2 = \gamma(A_3 - A_1 - A_2)$$

Here it is reasonable to consider the surface tension of the oil–water interface is a constant.

If we assume that the droplets are spherical, we have

$$\Delta F = F_{state-2} - F_{state-1} = 4\pi\gamma(a_3^2 - a_1^2 - a_2^2)$$

For simplicity, let us we assume $a_1 = a_2$,

Note $V_1 + V_2 = V_3$, $2a_1^3 = a_3^3$, and hence $2a_1^2 = \frac{a_3^3}{a_1}$.

$$\Delta F = 4\pi\gamma\left(a_3^2 - 2a_1^2\right) = 4\pi\gamma\left(a_3^2 - \frac{a_3^3}{a_1}\right) = 4\pi\gamma a_3^2(1 - \frac{a_3}{a_1})$$

Because

$$\frac{a_3}{a_1} > 1$$

$$\Delta F < 0$$

That is, this two-drop-merging process will reduce the total free energy of the system. Such a process will occur spontaneously.

Film Breakage and Droplet Formation on a Solid Surface

The breakage of a thin liquid film on a solid surface is an interfacial phenomenon that has important applications in industrial lubrication and coating processes. A thin liquid film on a solid surface may lose its stability and break into many discrete droplets on the solid surface. This interfacial phenomenon depends on the initial film thickness, the liquid surface tension and the contact angle. Within the framework of surface thermodynamics, this phenomenon can be modeled in terms of the free energy change of this film breaking-drop forming process.

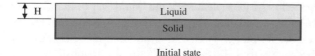

Initial state

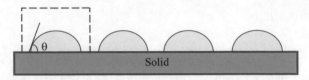

Final state

It can be shown that surrounding gas vapor phase and the solid phase will not contribute to this interfacial process. We will use the grand canonical free energy to model this process. The grand canonical free energy of the system at the initial state is given by:

$$\Omega_1 = -P_{L1}V_{L1} + \gamma_{lv}A_{lv1} + \gamma_{sl}A_{sl1}$$

where P_{L1} and V_{L1} are the pressure and the volume of the liquid film at the initial state, respectively; γ_{lv} and γ_{sl} are the surface tensions of the liquid–vapor interface and the solid–liquid interface, respectively; A_{lv1} and A_{sl1} are the surface areas of the liquid–vapor interface and the solid–liquid interface, respectively.

It should be noted that

$$A_{lv1} = A_{sl1} = A_0$$

The grand canonical free energy of the system at the initial state can be written as:

$$\Omega_1 = -P_{L1}V_{L1} + (\gamma_{lv} + \gamma_{sl})A_0$$

In the final state, let us assume that there are n droplets formed on the solid surface and all droplets have the same size. The grand canonical free energy of the system at the final state is given by:

$$\Omega_2 = n[-P_{L2}V_{L2} + \gamma_{lv}A_{lv2} + \gamma_{sl}A_{sl2} + \gamma_{sv}A_{sv}]$$

where n is the number of identical droplets formed on the solid surface; P_{L2} and V_{L2} are the pressure and the volume of each liquid drop at the final state, respectively; γ_{lv}, γ_{sv} and γ_{sl} are the surface tensions of the liquid–vapor interface, the solid–vapor surface and the solid–liquid interface, respectively; A_{lv2} and A_{sl2} are the surface areas of the liquid–vapor interface and the solid–liquid interface of each drop, respectively; A_{sv} is the surface area of the solid–vapor surface around each drop within the dash-line boundary as indicated in the figure above.

There are the following constraints:

$$V_{L1} = A_0H$$

$$V_{L1} = nV_{L2}$$

$$A_0 = n(A_{sv} + A_{sl2})$$

or

$$nA_{sv} = A_0 - nA_{sl2}$$

The free energy change of this film-breaking-and-droplet-forming process can then be written as:

$$
\begin{aligned}
\Delta\Omega = \Omega_2 - \Omega_1 &= P_{L1}V_{L1} - nP_{L2}V_{L2} + \gamma_{lv}(nA_{lv2} - A_0) \\
&\quad + \gamma_{sl}(nA_{sl2} - A_0) + n\gamma_{sv}A_{sv} \\
&= -(P_{L2} - P_{L1})V_{L1} + \gamma_{lv}(nA_{lv2} - A_0) \\
&\quad + (\gamma_{sl} - \gamma_{sv})(nA_{sl2} - A_0)
\end{aligned}
$$

Using the two volume constraints, we have

$$n = A_0 H / V_{L2}$$

The free energy change becomes:

$$\Delta\Omega = -(P_{L2} - P_{L1})A_0 H + \gamma_{lv}\left(\frac{A_0 H}{V_{L2}}A_{lv2} - A_0\right) + (\gamma_{sl} - \gamma_{sv})\left(\frac{A_0 H}{V_{L2}}A_{sl2} - A_0\right)$$

On a unit solid surface area basis,

$$\frac{\Delta\Omega}{A_0} = -(P_{L2} - P_{L1})H + \gamma_{lv}\left(\frac{H}{V_{L2}}A_{lv2} - 1\right) + (\gamma_{sl} - \gamma_{sv})\left(\frac{H}{V_{L2}}A_{sl2} - 1\right)$$

In the initial state, the pressure of the flat thin liquid film should be the same as the pressure of the vapor phase, that is,

$$P_{L1} = P_V$$

Assume all droplets are spherical and have a radius of R. From the Laplace equation of capillarity, we have

$$(P_{L2} - P_{L1}) = (P_{L2} - P_V) = \frac{2}{R}\gamma_{lv}$$

From the Young equation, we have

$$(\gamma_{sv} - \gamma_{sl}) = \gamma_{lv}cos\theta$$

Considering a droplet on the solid surface as a spherical cap,

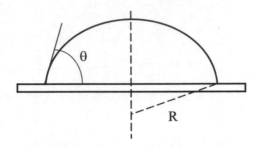

we have the following geometrical relations:

$$A_{sl2} = \pi R^2 sin^2\theta$$

$$A_{lv2} = 2\pi R^2(1 - cos\theta)$$

$$V_{L2} = \frac{\pi R^3}{3}(1 - cos\theta)^2(2 + cos\theta)$$

The free energy change can be re-written as:

$$\frac{\Delta\Omega}{A_0\gamma_{lv}} = -\frac{2}{R}H + \left(\frac{6}{R(1 - \cos\theta)(2 + \cos\theta)}H - 1\right)$$
$$- \cos\theta\left(\frac{3\sin^2\theta}{R(1 - \cos\theta)^2(2 + \cos\theta)}H - 1\right)$$

Clearly, the above equation indicates that the free energy change of such a process is a function of the initial film thickness, H, the droplet radius, R, and the contact angle, θ. Thermodynamically, if the free energy change is negative, such a film-breakage-and-droplet-form process is favourable (Fig. 3.1).

As we can see from the figure above, the free energy change initially decreases quickly with the radius of the formed droplets, and gradually reaches a constant value when the radius is larger than 50 μm. For the two cases where the contact angles are 10° and 30°, respectively, the free energy change is always positive. This is because the small contact angles in these two cases indicate very good wettability of the liquid to the solid surface, and hence the liquid tends to spread over the solid surface as a thin liquid film. In other words, the film breaking and forming droplets will increase the total free energy of the system, and it not a spontaneous process or not thermodynamically favorable process. However, for the case of 60° contact angle, the liquid does not wet the solid surface very well. If the thin liquid film breaks

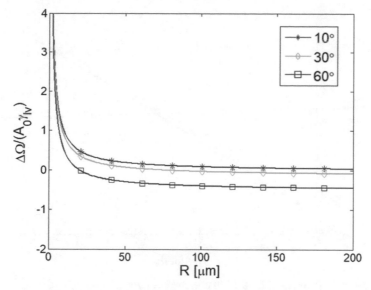

Fig. 3.1 Plot of $\frac{\Delta\Omega}{A_0\gamma_{lv}}$ versus R for $\theta = 10^\circ, 30^\circ, 60^\circ$, and $H = 10\,\mu m$

and form droplets with a radius larger than 25 μm, the free energy change becomes negative. This means that he film breaking and forming relatively larger droplets is thermodynamically favorable and may occur spontaneously (Fig. 3.2).

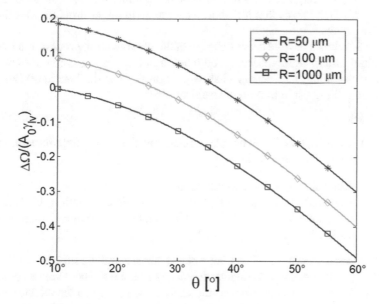

Fig. 3.2 Plot of $\frac{\Delta\Omega}{A_0\gamma_{lv}}$ versus θ for $R = 50, 100, 1000\,\mu m$, and $H = 10\,\mu m$

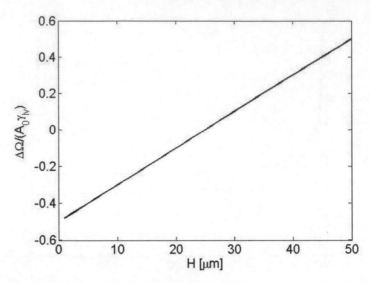

Fig. 3.3 Plot of $\frac{\Delta\Omega}{A_0\gamma_{lv}}$ versus H for R = 50 μm, and θ = 60°

As seen from the above figure, the free energy change is positive or zero when the contact angle is small (good wettability of the liquid), and becomes negative when the contact angle is larger (i.e., wettability decreases). For the same contact angle, for example, θ = 30°, the free energy change changes from a positive value for small droplets (e.g., R = 50 μm) to a negative value for large droplets (e.g., R = 1000 μm). This means that it is more likely to form larger droplets after the thin liquid film breaks.

Figure 3.3 shows the free energy change of the film breaking process as a function of the initial film thickness. Clearly, a thinner liquid film is preferable for breaking, as the corresponding free energy change is negative; and thicker liquid film most likely remains as a stable film on the solid surface.

Home Work

(1)   Model and analyze if two bubbles of different sizes can merge into one bubble in a bulk water phase. Under

   (a)   Constant gas volume assumption.
   (b)   Constant gas mass assumption. Why is this assumption different from the above assumption? Physically, which assumption is more close to reality?

(2)   Consider a process of oil wetting a planar surface and the surface is initially in contact with air. The oil contains no micro air bubbles. There are two possibilities: In the first case, the oil wets the surface completely. In the second case, the oil does not wet the surface completely and a bubble is formed on the surface. The following two figures illustrate the two possible scenarios. Assume that the

temperature and the pressures of the oil-air–solid surface system are the same for both cases. Which case is thermodynamically possible? (Hint: compare the Helmholtz free energies of these two cases).

Scenario 1   Scenario 2

Chapter 4
Second Law of Engineering Thermodynamics

Abstract In standard engineering thermodynamics textbooks, the core of thermodynamics is the first law of thermodynamics and the second law of thermodynamics. However, comprehension of the second law of thermodynamics is not simple. Particularly, the second law of thermodynamics is usually introduced to readers as the Kelvin-Planck statement of the second law and the Clausius statement of the second law. For students learning thermodynamics, the true meaning of these statements may not be easy to understand. Apparently, these two statements are very different. Why are both of them called the statement of the second law of thermodynamics? What do these statements have to do with irreversibility and entropy? In this chapter, we will answer these questions by providing analytical proof to show that the Kelvin-Planck statement and the Clausius statement of the second law are the results of the same thermodynamic principle—the second law (or the entropy increase principle), and how they are related to the irreversibility and the entropy generation.

In standard engineering thermodynamics textbooks, the core of thermodynamics is the first law of thermodynamics and the second law of thermodynamics. While the first law of thermodynamics, as the energy conservation principle, is easy to understand, comprehension of the second law of thermodynamics is not simple. Particularly, the second law of thermodynamics is introduced to readers as the Kelvin-Planck statement of the second law and/or the Clausius statement of the second law. The Kelvin-Planck statement of the second law is often articulated as:

It is impossible that a thermal cycle device receives heat from a single reservoir and produces a net amount of work.

The Clausius statement of the second law is often expressed as:

It is impossible to build a thermal cycle device that transfers heat from a lower-temperature body to a higher-temperature body without any work input.

For students learning thermodynamics for the first time, the meaning of these statements may not be easy to understand. On the surface, these two statements are very different, and talks about different things. Why are both of them called the statement of the second law of thermodynamics? Particularly, a key question may be: What do these statements have to do with irreversibility and entropy?

© The Author(s), under exclusive license to Springer Nature Switzerland AG 2022   177
D. Li, *Analytical Thermodynamics*,
https://doi.org/10.1007/978-3-030-90517-0_4

In addition, in the studies of thermal cycles, students may ask: In a thermal cycle, why do we have to reject heat to a heat sink? If there is no such a heat rejection, the thermal cycle would convert all heat supply into work and the efficiency would be 100%. Would this be possible?

In this chapter, we shall try to answer these questions and provide analytical proof to show that the Kelvin-Planck statement and the Clausius statement of the second law are the results of the same thermodynamic principle—the second law, and how they are related to the irreversibility and the entropy generation.

4.1 Irreversible Processes

In engineering thermodynamics, the interest is often about the performance of processes. A process is the change of thermodynamic state, i.e., from state 1 to state 2, for a given system. One would like to know, for example, what process can produce the maximum work for a given energy supply. However, in the studies of thermodynamic processes, one must realize that **a natural or spontaneous process moves only in a certain direction and cannot be reversed by itself, i.e., it is irreversible**. Real processes in our daily life and in engineering practice are irreversible. For example,

- Water flows down a waterfall.
- Gases expand from a high pressure place to a low pressure place.

- Heat flows from a high temperature place to a low temperature place.
- Time passes, we are getting older.

Furthermore, it should be realized that **a spontaneous process can be reversed; but, it will not be reversed by itself spontaneously. External input and energy must be provided to reverse the process**. For example, a stone is released from a certain height, and it falls to the ground. This is an irreversible process and the stone will not move back to its original position spontaneously. To bring it back (to reverse the process), one has to do work to pick it up. As another example, after heat is transferred from a high temperature body to a low temperature body, the heat will not flow back spontaneously from the low temperature body to the high temperature body. The heat transfer is an irreversible process. In order to make heat transfer back from the low temperature body to the high temperature body, one has to use a refrigerator or a heat pump. However, this requires using additional energy, for example, electricity to run the refrigerator or heat pump.

In order to understand the fact that all real processes are irreversible, let us also define the reversible process. **A reversible process is an idealized process** (which does not exist in reality). A reversible process is defined as a process **that can be reversed without leaving any impact on the surroundings**. This means, at the end of the reverse process, both the system and the surroundings are returned to their initial states. There is nothing changed in the world. From thermodynamics point of view, this is possible only if the net heat, network and mass exchange between the system and the surroundings are zero for the combined processes (i.e., the original process and the reverse process).

Because there are no reversible processes in real life, one cannot give you an example of a reversible process. However, one can use some examples of irreversible processes to **explain the meaning of the impact** on the surroundings after an irreversible process.

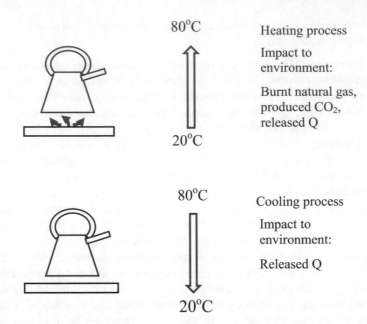

One example is the processes of heating and cooling water. Consider using a natural gas oven to heat a kettle of water from 20 to 80 °C. This is a heating process. Then, let the kettle of water to cool in the room from 80 to 20 °C. This is a cooling process to reverse the heating process. Now the system (the kettle of water) is recovered to its initial state. Is this a reversible process? No, because the surroundings have changed significantly by these processes. Let us examine what impact of these processes has left to the surroundings. (1) A certain amount of natural gas was burnt with air, some new gases were produced in the air by the combustion, such as carbon dioxide and water vapor. (2) A certain amount of heat was released into the air (from the natural gas combustion and from cooling the hot water), and may raise the temperature of the air in the room. Therefore, by definition, this (heating water) is not a reversible process.

4.2 Limitation of First Law of Thermodynamics

The **first law** of thermodynamics is the principle of energy conservation. It states that **energy cannot be created and cannot be destroyed; it can only change forms**. Now we understand that all real processes are irreversible. However, the first law of thermodynamics does not consider the irreversibility of the real processes, and therefore has shortcomings. For example:

(1)   As discussed above, **spontaneous processes can proceed only in a particular direction. The first law of thermodynamics gives no information about the**

direction of a spontaneous process. First law states only that when energy is transferred or converted from one form into another form, the quantity of energy must remain the same. We know by experience that heat flows spontaneously from a high temperature place to a low temperature place. But, heat transfer from a low temperature place to a higher temperature place without expenditure of external energy input to make the process take place would not violate the first law.

For example, the above figure shows that 1 kJ heat transfers out of the high temperature (T_H) body, and the low temperature (T_L) body receives 1 kJ heat. This is a spontaneous process and satisfies the first law of thermodynamics (i.e., the quantity of heat is conserved).

The above figure shows the reversed process. 1 kJ heat transfers out of the low temperature (T_L) body, and the high temperature (T_H) body receives 1 kJ heat. This process satisfies the first law of thermodynamics (i.e., the quantity of heat is conserved). However, this is not a spontaneous process and cannot occur naturally.

In other words, the first law of thermodynamics does not give any information on the permissible direction of a spontaneous process.

(2) **The first law is concerned with conservation of energy during energy transformation from one form to another. However, this also depends on the directions of energy transformation.** Joule's experiments showed that energy in the form of heat cannot be completely converted into work; however, work (e.g., electricity) energy can be completely converted into heat (thermal energy). Evidently, heat and work are not completely interchangeable forms of energy.

(3) **The 1st law efficiency analysis often fails to evaluate the true efficiency.** For example, the first law efficiency for a thermal cycle is defined by

$$\eta_I = \frac{W}{Q_H}$$

In this equation, the efficiency is determined only by the quantity of energy, i.e., the produced work (W) and the heat supply (Q_H), and the efficiency does not depend on the quality of the energy supply Q_H. However, the quality of the energy supply Q_H is a key to determine the true efficiency. For example, if a power-generation thermal cycle is provided with an amount of heat at a higher temperature, say 1000 K. Such a heat can produce superheated vapor to drive the turbine and consequently produce a lot of work and the efficiency of the thermal cycle is high. However, if the same thermal cycle is provided with the same amount of heat, but at a lower temperature, say 350 K (~77 °C), the heat at this temperature cannot even boil water; of course, the thermal cycle produces zero work and hence the efficiency is zero. Clearly, **we cannot find the true efficiency from a simple 1st law analysis**.

(4)   **When energy is transferred from one form to another, a part of the supplied energy always degrades into a less "useful" form**. For example, in a steam power plant, a part of the high temperature energy supplied by combustion is converted into useful work, but a significant part of the supplied heat is rejected to the atmosphere at a temperature close to atmospheric temperature. This rejected heat has a low temperature, and is essentially impossible to be used to generate any useful work (e.g., cannot boil water to generate vapor to rotate the turbine!). Therefore, **the quality (potential to produce work) of this part of the input energy is degraded during the work production process**. This is the **energy quality loss** due to the irreversibility of the process. 1st law ignores energy's quality.

It is important to recognize the quality of energy (not only the amount of energy). For heat, the higher the temperature, the higher the quality. Low quality or zero quality energy is useless to us. For example, we have practically infinite amount of energy stored in the environment surrounding us at $T \sim 300$ K and $P \sim 1$ atm. The problem is that, in order to utilize this energy source to produce useful work, we have to have another energy sink that is at a lower temperature, $T < 300$ K and/or a lower pressure, $P < 1$ atm. Unfortunately, such a "sink" does not exist naturally or not accessible. Deep space has essentially zero temperature and zero pressure; however, it is not accessible to us. A lower temperature space on earth can be created at the expenses of work and electricity (such as using a refrigerator). **Therefore, the environment around us at $T \sim 300$ K and $P \sim 1$ atm is a "dead state"**. The energy in this environment or the dead state has zero quality, or zero potential to produce any useful work and hence useless. We may **use this environment at $T_0 \sim 300$ K and $P_0 \sim 1$ atm as a mutual reference to evaluate the quality or work potential of any energy systems**. We will use T_0 and P_0 to represent the environmental temperature and pressure.

As discussed above, real processes are irreversible, and the first law of thermodynamics does not consider the irreversibility. Therefore, we need another law—the second law of thermodynamics to consider the irreversibility effects. **The irreversibility of a process is measured in terms of a thermodynamic property—Entropy S**.

4.3   Second Law and Equations

For real processes, there is the **principle of entropy increase**. It is stated as:

The entropy of an isolated system during a process always increases because of irreversibility. In the limiting case of a reversible process, entropy remains constant.

This is the **second law of thermodynamics**. Mathematically, it is expressed by:

$$\Delta S_{\text{isolated system}} = S_{gen} \geq 0$$

where S_{gen} is the entropy generation of the system due to the irreversibility; "=" sign holds for the idealized reversible process; and ">" sign holds for the irreversible process.

Generally, for a non-isolated system (i.e., a system interacting with surroundings), the combined system, i.e., the system plus the surroundings, forms an isolated system. **The 2nd law can apply to the combined system as**:

$$\Delta S_{\text{isolated combined system}} = \Delta S_{sys} + \Delta S_{surr} = S_{gen} \geq 0$$

The above equation states that the total entropy change for a process is the amount of entropy generated during that process (S_{gen}), and it is equal to the sum of the entropy changes of the system and the surroundings. It is not difficult to understand that the entropy change of the system and the entropy change of the surroundings individually do not have to be positive, however, the sum of them (equals to S_{gen}) must be positive. That is, the **sum** of the entropy changes of the system and its surroundings for an isolated system can never decrease, because of entropy generated during that irreversible process, S_{gen}.

Let us further analyze the **2nd law equation**:

$$S_{gen} = \Delta S_{sys} + \Delta S_{surr}$$

Entropy change of a system

The entropy change of a system is the result of the process occurring within the system.

Entropy change of the system = Entropy at final state − Entropy at initial state

$$\Delta S_{system} = S_{final} - S_{initial} = S_2 - S_1$$

Entropy change of the surrounding

How to evaluate ΔS_{surr} (or what is ΔS_{surr})? Let's look at the entropy balance in a process as illustrated in the figure below.

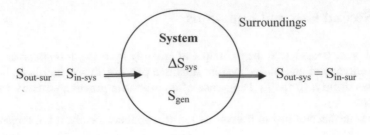

By inspecting the figure above, we have:

$$\Delta S_{surr} = S_{in-sur} - S_{out-sur} = S_{out-sys} - S_{in-sys} = (S_{out} - S_{in})_{sys}$$

Therefore, the 2nd law equation now can be written as:

$$S_{gen} = \quad \Delta S_{sys} \quad + \quad \underbrace{(S_{out} - S_{in})_{sys}} \qquad (kJ/K)$$

$\qquad\qquad (S_2 - S_1)$ entropy transfer across the system boundary

For a system with multiple boundaries, the 2nd law equation can be written as:

$$S_{gen} \quad = \quad \Delta S_{sys} \quad + \quad \underbrace{\sum (S_{out} - S_{in})_{sys}} \qquad (kJ/K)$$

Entropy	Entropy change	Net entropy transfer
generation	of the system	across system's boundaries

It can also be expressed in the general **time rate form** *(where the dot on top of each symbol indicates the time rate, i.e., change per unit time)* as:

$$\dot{S}_{gen} \quad = \quad \Delta \dot{S}_{sys} \quad + \quad \sum (\dot{S}_{out} - \dot{S}_{in})_{sys} \qquad (kW/K)$$

rate of entropy	rate of entropy	rate of net entropy transfer
generation	change	across system's boundaries
	of the system	

Entropy transfer across the boundary

How to evaluate the entropy transfer across the system boundaries? Entropy can be transferred to or from a system by two mechanisms: heat transfer and mass flow.

Entropy transfer with mass flow

Mass carries entropy as well as energy, and the total entropy and the total energy contents of a system are proportional to the mass. When a certain amount mass enters or leaves a system, correspondingly, entropy associated with this mass also enters or leaves the system. Let $\dot{m}$ be the mass flow rate (kilogram per second, i.e., kg/s) and s be the specific entropy of the mass (kJ/kgK), the time rate of entropy transfer with mass transfer is given by:

$$\dot{s}_{mass} = \dot{m}s$$

Entropy transfer associated with heat transfer

The entropy transfer associated with heat transfer is defined by the ratio of the heat transfer Q at a location to the absolute temperature T at the same location, and is given by:

$$S_{Heat} = S_Q = \frac{Q}{T}$$

where the temperature T is constant.

As indicated by the above equation, the direction of entropy transfer is the same as the direction of heat transfer since the absolute temperature T is always a positive quantity.

If the system has multiple boundaries and each boundary has a different temperature, the entropy transfer associated with heat transfer across these boundaries can be evaluated by:

$$S_Q = \sum \frac{Q_k}{T_k}$$

When the temperature at the boundary is not constant, the entropy transfer associated with the heat transfer across the boundary can be determined by integration as:

$$S_Q = \int_1^2 \frac{dQ}{T}$$

2nd Law equation for closed systems

For a closed system, by definition, there is no mass transfer across the system boundary. The entropy transfer term for a closed system, $\sum(S_{out} - S_{in})$, in the 2nd law equation will be associated with heat transfer only. The 2nd law equation for closed systems is:

$$S_{gen} = (S_2 - S_1)_{sys} + \underbrace{\sum \frac{Q_{j,out}}{T_{j,out}} - \sum \frac{Q_{k,in}}{T_{k,in}}}_{\sum (S_{out} - S_{in})_{sys}} \quad (kJ/k)$$

entropy transfer by heat transfer

2nd Law equation for open systems

The 2nd law equation for open systems differs from that for closed systems because the entropy exchange due to mass flow must be included. In the time rate form we have

$$\dot{S}_{gen} = \Delta \dot{S}_{sys} + \underbrace{\sum \dot{m}_{out} s_{out} - \sum \dot{m}_{in} s_{in} + \sum \frac{\dot{Q}_{j,out}}{T_{j,out}} - \sum \frac{\dot{Q}_{k,in}}{T_{k,in}}}_{\sum (\dot{S}_{out} - \dot{S}_{in})_{sys}} \quad (kW/k)$$

Entropy transfer rate with mass transfer and heat transfer

For a **steady-state, steady-flow process**, by definition, the system's properties are independent of time, therefore,
$\Delta \dot{S}_{sys} = 0$.
The second law equation is reduced to:

$$\dot{S}_{gen} = \sum \dot{m}_{out} s_{out} - \sum \dot{m}_{in} s_{in} + \sum \frac{\dot{Q}_{j,out}}{T_{j,out}} - \sum \frac{\dot{Q}_{k,in}}{T_{k,in}} \quad (kW/k)$$

For a single-stream (one inlet and one exit) **steady-flow** device, $\dot{m}_{out} = \dot{m}_{in} = \dot{m}$. The second law equation becomes

$$\dot{S}_{gen} = \dot{m}(s_{out} - s_{in}) + \sum \frac{\dot{Q}_{j,out}}{T_{j,out}} - \sum \frac{\dot{Q}_{k,in}}{T_{k,in}} \quad (kW/k)$$

4.4 The 2nd Law Requirement on Thermal Cycles

From the above analysis, we conclude the following:

(1) The 1st law of thermodynamics requires that energy must be conserved in a process.
(2) The 2nd law of thermodynamics states that the entropy generation must be positive in a real process.
(3) **A process cannot take place unless it satisfies both the 1st law and the 2nd law.**

Apply both the 1st law and the 2nd law to analyze a thermal engine cycle.

Consider a thermal engine cycle system working between two heat sources, as illustrated in the figure below.

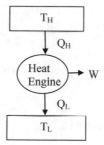

In the above figure, the heat engine is a device that operates in a thermal cycle. During the thermal cycle, the heat engine receives heat (Q_H) from a high temperature (T_H) heat reservoir, converts a part of the heat supply into work (W), and rejects the rest of the heat (Q_L) to a low temperature (T_L) heat reservoir (heat sink).

A thermal cycle is a closed system because the mass of the working fluid in the thermal cycle is constant. The 1st law equation for the thermal cycle as a closed system is given by:

$$Q_{net} - W_{net} = \Delta U$$

Because the starting state and the ending state are identical in a cycle, and hence no change in the properties of the system on per cycle basis, therefore,

$$\Delta U = 0 \text{ (per cycle)}$$

and

$$Q_{net} = Q_{in} - Q_{out} = Q_H - Q_L$$

Therefore, the 1st law equation becomes:

$$W_{net} = Q_H - Q_L \tag{4.1}$$

The 2nd law equation for the thermal cycle as a closed system is given by:

$$S_{gen} = \Delta S_{cycle} + \sum \frac{Q_{k,out}}{T_{k,out}} - \sum \frac{Q_{j,in}}{T_{j,in}} \; (kJ/K)$$

Again, because the starting state and the ending state are identical in a cycle, and hence no change in the properties of the system on per cycle basis, therefore,

$$\Delta S_{cycle} = 0 \; (per\,cycle)$$

And note that in this thermal cycle,

$$Q_{in} = Q_H \text{ and } Q_{out} = Q_L$$

the 2nd law equation becomes:

$$S_{gen} = \frac{Q_L}{T_L} - \frac{Q_H}{T_H} \tag{4.2}$$

Multiplying the 2nd law equation, Eq. (4.2), by the environmental temperature T_0,

$$T_0 S_{gen} = \frac{T_0}{T_L} Q_L - \frac{T_0}{T_H} Q_H$$

and combining this equation with the 1st law equation, Eq. (4.1), yields

$$W_{net} + T_0 S_{gen} = Q_H - Q_L + \frac{T_0}{T_L} Q_L - \frac{T_0}{T_H} Q_H$$

It can be rearranged into:

$$W_{net} = Q_H \left[1 - \frac{T_0}{T_H}\right] - Q_L \left[1 - \frac{T_0}{T_L}\right] - T_0 S_{gen} \tag{4.3}$$

The 1st term on the right-hand side of Eq. (4.3) is the maximum work that can be produced from the heat supply Q_H. The 2nd term on the right-hand side of Eq. (4.3) is the potential work associated with the rejected heat Q_L (if $T_L > T_0$). The last term on the right-hand side is the work potential destroyed by the irreversibility involved in the cycle.

The above equation can be further rearranged in the following form:

$$T_0 S_{gen} = Q_H \left[1 - \frac{T_0}{T_H}\right] - Q_L \left[1 - \frac{T_0}{T_L}\right] - W_{net} \tag{4.4}$$

- **Discussion of special case 1:**

Assume $Q_L = 0$, that is, no heat rejection to the heat sink.

$$\eta_{th} = \frac{W}{Q_H} = 1 - \frac{Q_L}{Q_H} = 1 = 100\%$$ Is it possible?

Under this assumption, the 1st law becomes:

$$W_{net} = Q_H - \cancel{Q_L} = Q_H.$$

If so, from Eq. (4.3), we have:

$$T_0 S_{gen} = Q_H \left[1 - \frac{T_0}{T_H} \right] - W_{net}$$

$$= Q_H \left[1 - \frac{T_0}{T_H} \right] - Q_H$$

$$= Q_H \left[-\frac{T_0}{T_H} \right] < 0$$

This implies $S_{gen} < 0$, and violates the 2nd law. Hence such a heat engine is not possible. This is the **Kelvin-Planck Statement of the Second Law**:
   It is impossible that a thermal cycle device receives heat from a single source and produce a net amount of work.

The Kelvin-Planck statement of the second law of thermodynamics states that no heat engine can produce a net amount of work while exchanging heat with a single reservoir only. There must be heat rejection to a low temperature heat sink. In other words, the maximum possible efficiency is always less than 100%.

Apply both the 1st law and the 2nd law to analyze a refrigeration cycle.

As illustrated in the figure below, a refrigerator or a heat pump is a device that operates in a thermal cycle. During the thermal cycle, the refrigerator receives a work input (W), extracts a certain amount of heat (Q_L) from a low temperature (T_L) heat reservoir (heat sink), and rejects heat (Q_H) to a high temperature (T_H) heat reservoir.
   For a refrigeration cycle, applying the 1st law and the 2nd law, we can derive an equation similar to that (Eq. (4.3)) of the thermal engine. The 1st law equation for a thermal cycle is given by:

$$Q_{net} - W_{net} = \Delta U$$

As explained before, because the starting state and the ending state are identical in a cycle, and hence no change in the properties of the system on per cycle basis,

therefore,

$$\Delta U = 0 \,(\text{per cycle})$$

The first law equation is reduced to.

$$Q_{net} - W_{net} = 0.$$

For a refrigeration cycle,

$$W_{net} = W_{\text{done by system}} - W_{\text{done to system}}$$
$$= W_{\text{out}} - W_{\text{in}} = -W_{\text{in}}$$

and

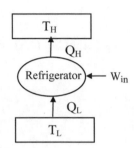

$$Q_{net} = Q_{in} - Q_{out} = Q_L - Q_H$$

so that the 1st law equation becomes

$$-W_{in} = Q_L - Q_H \qquad (4.1a)$$

The 2nd law equation for the thermal cycle as a closed system is given by:

$$S_{gen} = \Delta S_{cycle} + \sum \frac{Q_{k,out}}{T_{k,out}} - \sum \frac{Q_{j,in}}{T_{j,in}} \,(\text{kJ/K})$$

Again, because the starting state and the ending state are identical in a cycle, and hence no change in the properties of the system on per cycle basis, therefore,

$$\Delta S_{cycle} = 0$$

(cycle),

and for the refrigeration cycle, as shown in the figure above,

$$Q_{out} = Q_H \text{and} Q_{in} = Q_L,$$

The second law equation becomes:

$$S_{gen} = \frac{Q_H}{T_H} - \frac{Q_L}{T_L} \qquad (4.2a)$$

Multiplying the 2nd law equation, Eq. (4.2a), by the environmental temperature T_0 yields:

$$T_0 S_{gen} = \frac{T_0}{T_H} Q_H - \frac{T_0}{T_L} Q_L$$

combining this equation with the 1st law equation, Eq. (4.1a), leads to:

$$T_0 S_{gen} = Q_L \left[1 - \frac{T_0}{T_L}\right] - Q_H \left[1 - \frac{T_0}{T_H}\right] + W_{in} \qquad (4.3a)$$

- **Discussion of special case 2:**

Assume $W_{in} = 0$, that is, no external work input to run the refrigerator. This implies that the COP (Coefficient of Performance) of a refrigerator or heat pump is infinity.

$$COP_R = \frac{Q_L}{W_{in}} = \frac{Q_L}{0} = \infty \qquad \text{Is this possible?}$$

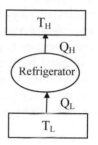

In this case, the 1st law equation, Eq. (4.1a), becomes:

$$Q_L = Q_H = Q.$$

From Eq. (4.3a), we have:

$$\begin{aligned} T_0 S_{gen} &= Q_L \left[1 - \frac{T_0}{T_L}\right] - Q_H \left[1 - \frac{T_0}{T_H}\right] \\ &= Q \left[1 - \frac{T_0}{T_L} - 1 + \frac{T_0}{T_H}\right] \\ &= Q T_0 \left[\frac{1}{T_H} - \frac{1}{T_L}\right] < 0 \end{aligned}$$

This implies $S_{gen} < 0$, and violates the 2nd law. Hence such a refrigerator cannot exist. This is the **Clausius Statement of the Second Law**:

It is impossible to construct a heat pump device that can transfer heat from a low temperature place to a high temperature place without any work input.

Thus, the COP of a refrigerator or heat pump cannot be infinity.

4.5 Applying Second Law Equation to a Human Body

Thinking about the 2nd law equation for open systems, one may ask if we could apply the 2nd law equation to a living biological system such as a cell or a human body. A human body, for example, is an open system that exchanges mass (e.g., air, water and food) and energy (heat) with the surroundings. Therefore, the 2nd law equation should be applicable.

$$\dot{S}_{gen} = \Delta\dot{S}_{sys} + \sum \dot{m}_{out}s_{out} - \sum \dot{m}_{in}s_{in} + \sum \frac{\dot{Q}_{j,out}}{T_{j,out}} - \sum \frac{\dot{Q}_{k,in}}{T_{k,in}} \quad (kW/k)$$

Obviously, all the bio-chemical processes taking place inside our body are irreversible. Without the need to know the specifics of each of these processes, we know that, eventually, the net effect of these irreversible processes is to make us aging (getting old), to make the functions and the performance of the body parts and organs deteriorating, and resulting in diseases. Therefore, from thermodynamics point of view, the body's health condition is determined by the net effect of the internal irreversibility of a human body and should be reflected by the total entropy generation rate. The value of this entropy generation rate should indicate the speed of the internal irreversibility. Of course, everyone likes to have a long and healthy life. This will require having a low entropy generation rate or a minimum entropy generation rate under given conditions. Obviously, a lot of research has to be done to understand these internal irreversible processes and find the controlling parameters to reduce the total entropy generation rate. One may imagine that someday in the future when you go to visit your family doctor, the first thing the doctor would say to you is: Ok, let me measure your entropy generation rate first. (Of course, the medical students then would have to learn thermodynamics before they could become doctors).

Generally, the temperature of a human body is 37 °C, higher than the average environmental temperature. Therefore, there is no heat transfer from the surrounding to the human body. That is, **The equation below should be placed in the center.

$$\sum \frac{\dot{Q}_{k,in}}{T_{k,in}} = 0$$

On the other hand, heat transfers from human body to the surrounding. We have. **The equation below should be placed in the center.

$$\sum \frac{\dot{Q}_{j,out}}{T_{j,out}} = \frac{\dot{Q}_{out}}{T_{body}}$$

Furthermore, within a short period of time, we may consider the body is in a steady state, and hence $\Delta \dot{S}_{sys} = 0$.

The mass transfer into the body generally includes air, food and drinks. The mass transfer out of the body includes the rejected wastes and moisture vaporized from the skin. Therefore, the 2nd law equation for an average human body becomes:

$$\dot{S}_{gen} = \sum \dot{m}_{out} s_{out} - \sum \dot{m}_{in} s_{in} + \frac{\dot{Q}_{out}}{T_{body}} \quad (kW/K)$$

It is pretty amazing to see that such a simple equation could allow us to estimate the internal irreversibility of a human body. Of course, this is a phenomenological equation; it calculates the entropy generation by using the "outside" or "apparent" measurable parameters (not by the parameters characterizing the internal irreversible processes).

Of course, we may apply the same approach to individual processes or segments of a human body, to analyze the entropy generation rate or the irreversibility of these individual processes or segments. Such analysis would tell us the key and problematic segments of the body, pointing out the directions to improve our health and life.

In the above 2nd law equation, the entropy rejected from human body with mass is given by

$$+ \sum \dot{m}_{out} s_{out}$$

and the entropy rejected from human body with heat is given by

$$+ \frac{\dot{Q}_{out}}{T_{body}}$$

These are two positive terms in the above 2nd law equation. Clearly, they indicate that we release positive entropy. In other words, our body is like an entropy generation machine because of the internal irreversibility.

Intuitively, we know we have to breath fresh air, eat food and drink water to maintain our live. Air, food and drinks all have entropies. From the 2nd law equation given above, we see clearly that these intake entropies are represented by

$$- \sum \dot{m}_{in} s_{in}$$

which is a negative term in the above 2nd law equation. This means that the entropy we take into our body is "negative". In other words, we take "negative entropy" to balance some of the positive entropies generated by the body in order to maintain our life. Perhaps, tomorrow morning at the breakfast table, instead of asking for a cup of milk, you should say, "A cup of negative entropy, please."

Printed in the United States
by Baker & Taylor Publisher Services